VALVE SELECTION HANDBOOK

FIFTH EDITION

T0343350

V A L V E
S E L E C T I O N
H A N D B O O K

FIFTH EDITION

Engineering fundamentals for selecting the right
valve design for every industrial flow application

By

P E T E R   S M I T H
(Editor and Contributor)

and

R .   W .   Z A P P E

AMSTERDAM • BOSTON • LONDON • NEW YORK • OXFORD • PARIS
SAN DIEGO • SAN FRANCISCO • SINGAPORE • SYDNEY • TOKYO

Gulf Professional Publishing is an imprint of Elsevier Inc.

ELSEVIER

G | P
P |

Gulf Professional Publishing is an imprint of Elsevier
200 Wheeler Road, Burlington, MA 01803, USA
Linacre House, Jordan Hill, Oxford OX2 8DP, UK

Recognizing the importance of preserving what has been written, Elsevier prints its books on acid-free paper whenever possible.

Library of Congress Cataloging-in-Publication Data
Application submitted

British Library Cataloguing-in-Publication Data
A catalogue record for this book is available from the British Library.

ISBN: 0-7506-7717-1

For information on all Gulf Professional Publishing
publications visit our website at www.gulfpp.com

Transferred to Digital Printing in 2011

CONTENTS

Flow Through Valves, 28

Resistance Coefficient ζ, 29. Flow Coefficient C_v, 33.
Flow Coefficient K_v, 34. Flow Coefficient A_v, 35.
Interrelationships Between Resistance and Flow Coefficients, 36.
Relationship Between Resistance Coefficient and Valve Opening
Position, 37. Cavitation of Valves, 38. Waterhammer from
Valve Operation, 40. Attenuation of Valve Noise, 44.

3 Manual Valves 47

Functions of Manual Valves, 47

Grouping of Valves By Method of Flow Regulation, 47

Selection of Valves, 48

Valves for Stopping and Starting Flow, 48. Valves for
Controlling Flow Rate, 48. Valves for Diverting Flow, 49.
Valves for Fluids with Solids in Suspension, 50. Valve End
Connections, 50. Standards Pertaining to Valve Ends, 51.
Valve Ratings, 51. Valve Selection Chart, 52.

Globe Valves, 54

Valve Body Patterns, 54. Valve Seatings, 61.
Connection of Disc to Stem, 64. Inside and Outside Stem Screw, 64.
Bonnet Joints, 65. Stuffing Boxes and Back Seating, 66.
Direction of Flow Through Globe Valves, 68.
Standards Pertaining to Globe Valves, 69. Applications, 69.

Piston Valves, 69

Construction, 72. Applications, 73.

Parallel Gate Valves, 73

Conventional Parallel Gate Valves, 76. Conduit Gate
Valves, 79. Valve Bypass, 82. Pressure-Equalizing
Connection, 82. Standards Pertaining to Parallel Gate Valves, 83.
Applications, 83.

Wedge Gate Valves, 84

Variations of Wedge Design, 87. Connection of Wedge to Stem, 92.
Wedge Guide Design, 93. Valve Bypass, 94. Pressure-Equalizing
Connection, 94. Case Study of Wedge Gate Valve Failure, 94.
Standards Pertaining to Wedge Gate Valves, 96.
Applications, 96.

Plug Valves, 97

Cylindrical Plug Valves, 98. Taper Plug Valves, 101.
Antistatic Device, 104. Plug Valves for Fire Exposure, 105.
Multiport Configuration, 106. Face-to-Face Dimensions and Valve
Patterns, 107. Standards Pertaining to Plug Valves, 107.
Applications, 108.

Ball Valves, 108

Seat Materials for Ball Valves, 109. Seating Designs, 109.
Pressure-Equalizing Connection, 114. Antistatic Device, 115.
Ball Valves for Fire Exposure, 115. Multiport Configuration, 117.
Ball Valves for Cryogenic Service, 117. Variations of Body
Construction, 117. Face-to-Face Dimensions, 119.
Standards Pertaining to Ball Valves, 119.
Applications, 119.

Butterfly Valves, 120

Seating Designs, 121. Butterfly Valves for Fire Exposure, 133.
Body Configurations, 133. Torque Characteristic of Butterfly
Valves, 134. Standards Pertaining to Butterfly Valves, 136.
Applications, 137.

Pinch Valves, 137

Open and Enclosed Pinch Valves, 139. Flow Control with
Mechanically Pinched Valves, 140. Flow Control with
Fluid-Pressure Operated Pinch Valves, 140. Valve Body, 141.
Limitations, 142. Standards Pertaining to Pinch Valves, 142.
Applications, 143.

Diaphragm Valves, 143

Weir-Type Diaphragm Valves, 144. Straight-Through
Diaphragm Valves, 145. Construction Materials, 146.
Valve Pressure/Temperature Relationships, 147. Valve Flow
Characteristics, 147. Operational Limitations, 149. Standards
Pertaining to Diaphragm Valves, 149.
Applications, 149.

Stainless Steel Valves, 149

Corrosion-Resistant Alloys, 149. Crevice Corrosion, 150.
Galling of Valve Parts, 150. Light-Weight Valve
Constructions, 151. Standards Pertaining to Stainless Steel
Valves, 151.

PREFACE

FIFTH EDITION

I originally purchased the *Valve Selection Handbook* first edition way back in 1982, during my early years in the oil and gas industry. I was then working in Indonesia for a company called Huffco, a U.S.-based independent operator, who had oil and gas facilities on the island of Borneo. Starved of information in this jungle environment and during the pre-Internet days, I needed a reference book that would give me some professional guidance. I found this information in the *Valve Selection Handbook*, which either resolved the current problem that I was facing or "turned on the lights" and pointed me in the right direction.

I am a great believer in the Internet for general searching, however I always prefer to review detailed information "off paper." Even the various software packages that allow you to search through a specific volume have their limitations. For this reason I still believe that reference books are a relevant method of presenting and accessing technical information and data. Word searches help, but sometimes you do not know exactly what you are looking for until you have found it.

We are now at edition five and, although valve engineering is far from rocket science, advances have been made over the last quarter century in design, manufacturing processes, and materials for construction. Materials previously known as "exotic," such as Duplex, Monel®, Inconel®, and titanium, are know used more commonly and have become economically viable solutions to engineering problems.

I was honored when I was invited by Phil Carmical of Elsevier Science to edit the fifth edition of this title, because it was a book that I not only owned, but also respected because of its usefulness. I have made very subtle changes to the original text and, because the original was concise and to the point, I have adopted the same philosophy for the additional chapters on actuators, double block and bleed ball valves, and valve locking devices that have been written for this fifth edition. I have also included a glossary of valve terminology that will assist the reader.

I value any feedback via the publisher on the content of this book and where appropriate it will be incorporated in the next edition.

Peter Smith
London, April 2003

1

INTRODUCTION

The purpose of this book is to assist the Piping Specifying Engineer in the selection of valves for a specific application and that meet the design parameters of the process service. Valve selection is based on function, material suitability, design pressure/temperature extremities, plant life, end connections, operation, weight, availability, maintenance, and cost. I have deliberately placed cost at the end for a reason. If the valve does not meet the design criteria, then even if it is free, it is still too expensive, because of the costs to replace it when it fails. Just like life, valve selection is a series of compromises.

Valves are the components in a fluid flow or pressure system that regulate either the flow or the pressure of the fluid. This duty may involve stopping and starting flow, controlling flow rate, diverting flow, preventing back flow, controlling pressure, or relieving pressure.

These duties are performed by adjusting the position of the closure member in the valve. This may be done either manually or automatically. Manual operation also includes the operation of the valve by means of a manually controlled power operator. The valves discussed here are manually operated valves for stopping and starting flow, controlling flow rate, and diverting flow; and automatically operated valves for preventing back flow and relieving pressure. The manually operated valves are referred to as manual valves, while valves for the prevention of back flow and the relief of pressure are referred to as check valves and pressure relief valves, respectively.

Rupture discs are non-reclosing pressure-relieving devices which fulfill a duty similar to pressure relief valves.

Fundamentals

Sealing performance and flow characteristics are important aspects in valve selection. An understanding of these aspects is helpful and often essential in the selection of the correct valve. Chapter 2 deals with the fundamentals of valve seals and flow through valves.

The discussion on valve seals begins with the definition of fluid tightness, followed by a description of the sealing mechanism and the design of seat seals, gasketed seals, and stem seals. The subject of flow through valves covers pressure loss, cavitation, waterhammer, and attenuation of valve noise.

Manual Valves

- Stopper type closure—globe, needle
- Vertical slide—gate
- Rotary type—ball, plug, butterfly
- Flexible body—diaphragm

Manual valves are divided into four groups according to the way the closure member moves onto the seat. Each valve group consists of a number of distinct types of valves that, in turn, are made in numerous variations.

The way the closure member moves onto the seat gives a particular group or type of valve a typical flow-control characteristic. This flow-control characteristic has been used to establish a preliminary chart for the selection of valves. The final valve selection may be made from the description of the various types of valves and their variations that follow that chart.

Note: For literature on **control valves**, refer to footnote on page 5 of this book.

Check Valves

- Lift check
- Swing check (single and double plate)

* Tilting disc
* Diaphragm

The many types of check valves are also divided into four groups according to the way the closure member moves onto the seat.

The basic duty of these valves is to prevent back flow. However, the valves should also close fast enough to prevent the formation of a significant reverse-flow velocity, which on sudden shut-off, may introduce an undesirably high surge pressure and/or cause heavy slamming of the closure member against the seat. In addition, the closure member should remain stable in the open valve position.

Chapter 4, on check valves, describes the design and operating characteristics of these valves and discusses the criteria upon which check valves should be selected.

Pressure Relief Valves

* Direct-loaded pressure relief valves
* Piloted pressure relief valves

Pressure relief valves are divided into two major groups: direct-acting pressure relief valves that are actuated directly by the pressure of the system fluid, and pilot-operated pressure relief valves in which a pilot controls the opening and closing of the main valve in response to the system pressure.

Direct-acting pressure may be provided with an auxiliary actuator that assists valve lift on valve opening and/or introduces a supplementary closing force on valve reseating. Lift assistance is intended to prevent valve chatter while supplementary valve loading is intended to reduce valve simmer. The auxiliary actuator is actuated by a foreign power source. Should the foreign power source fail, the valve will operate as a direct-acting pressure relief valve.

Pilot-operated pressure relief valves may be provided with a pilot that controls the opening and closing of the main valve directly by means of an internal mechanism. In an alternative type of pilot-operated pressure relief valve, the pilot controls the opening or closing of the main valve indirectly by means of the fluid being discharged from the pilot.

A third type of pressure relief valve is the powered pressure relief valve in which the pilot is operated by a foreign power source. This type of pressure relief valve is restricted to applications only that are required by code.

Rupture Discs

Rupture discs are non-reclosing pressure relief devices that may be used alone or in conjunction with pressure relief valves. The principal types of rupture discs are forward domed types, which fail in tension, and reverse buckling types, which fail in compression. Of these types, reverse buckling discs can be manufactured to close burst tolerances. On the debit side, not all reverse buckling discs are suitable for relieving incompressible fluids.

While the application of pressure relief valves is restricted to relieving nonviolent pressure excursions, rupture discs may be used also for relieving violent pressure excursions resulting from the deflagration of flammable gases and dust. Rupture discs for deflagration venting of atmospheric pressure containers or buildings are referred to as vent panels.

Units of Measurement

Measurements are given in SI and imperial units. Equations for solving in customary but incoherent units are presented separately for solution in SI and imperial units as presented customarily by U.S. manufacturers. Equations presented in coherent units are valid for solving in either SI or imperial units.

Identification of Valve Size and Pressure Class

The identification of valve sizes and pressure classes in this book follows the recommendations contained in MSS Standard Practice SP-86. Nominal valve sizes and pressure classes are expressed without the addition of units of measure; e.g., NPS 2, DN 50 and Class I 50, PN 20. NPS 2 stands for nominal pipe size 2 in. and DN 50 for diameter nominal 50 mm. Class 150 stands for class 150 lb. and PN 20 for pressure nominal 20 bar.

Standards

Appendix C contains the more important U.S., British, and ISO standards pertaining to valves. The standards are grouped according to valve type or group.

Additional Chapters

There are three additional chapters in the fifth edition of the *Valve Selection Handbook* that have not been included previously:

Chapter 8—Actuators
Chapter 9—Double Block and Bleed Ball Valves
Chapter 10—Mechanical Locking Devices for Valves

A comprehensive glossary has also been included in Appendix E to assist the reader.

This book does not deal with control valves. Readers interested in this field should consult the following publications of the ISA:

1. *Control Valve Primer, A User's Guide* (3rd edition, 1998), by H. D. Baumann. This book contains new material on valve sizing, smart (digital) valve positioners, field-based architecture, network system technology, and control loop performance evaluation.
2. *Control Valves, Practical Guides for Measuring and Control* (1st edition, 1998), edited by Guy Borden. This volume is part of the *Practical Guide Series*, which has been developed by the ISA. The last chapter of the book deals also with regulators and compares their performance against control valves. Within the *Practical Guide Series*, separate volumes address each of the important topics and give them comprehensive treatment.

Address: ISA, 67 Alexander Drive, Research Triangle Park, NC 27709, USA. Email http://www.isa.org

2

FUNDAMENTALS

The fundamentals of a particular type of valve relate to its sealing characteristics, which include in-line seat sealing when closed and where applicable stem sealing which should prevent potential leaks into the atmosphere. In the case of process systems handling hazardous fluids, harmful to both the atmosphere and personnel, stem sealing is considered to be of more importance.

FLUID TIGHTNESS OF VALVES

Valve Seals

One of the duties of most valves is to provide a fluid seal between the seat and the closure member. If the closure member is moved by a stem that penetrates into the pressure system from the outside, another fluid seal must be provided around the stem. Seals must also be provided between the pressure-retaining valve components. If the escape of fluid into the atmosphere cannot be tolerated, the latter seals can assume a higher importance than the seat seal. Thus, the construction of the valve seals can greatly influence the selection of valves.

Leakage Criterion

A seal is fluid-tight if the leakage is not noticed or if the amount of noticed leakage is permissible. The maximum permissible leakage for the application is known as the leakage criterion.

The fluid tightness may be expressed either as the time taken for a given mass or volume of fluid to pass through the leakage capillaries or as the time taken for a given pressure change in the fluid system. Fluid tightness is usually expressed in terms of its reciprocal, that is, leakage rate or pressure change.

Four broad classes of fluid tightness for valves can be distinguished: nominal-leakage class, low-leakage class, steam class, and atom class.

The nominal- and low-leakage classes apply only to the seats of valves that are not required to shut off tightly, as commonly in the case for the control of flow rate. Steam-class fluid tightness is relevant to the seat, stem, and body-joint seals of valves that are used for steam and most other industrial applications. Atom-class fluid tightness applies to situations in which an extremely high degree of fluid tightness is required, as in spacecraft and atomic power plant installations.

Lok[1] introduced the terms *steam class* and *atom class* for the fluid tightness of gasketed seals, and proposed the following leakage criteria.

Steam Class:
Gas leakage rate 10 to 100 μg/s per meter seal length.
Liquid leakage rate 0.1 to 1.0 μg/s per meter seal length.
Atom Class:
Gas leakage rate 10^{-3} to 10^{-5} μg/s per meter seal length.

In the United States, atom-class leakage is commonly referred to as zero leakage. A technical report of the Jet Propulsion Laboratory, California Institute of Technology, defines zero leakage for spacecraft requirements.[2] According to the report, zero leakage exists if surface tension prevents the entry of liquid into leakage capillaries. Zero gas leakage as such does not exist. Figure 2-1 shows an arbitrary curve constructed for the use as a specification standard for zero gas leakage.

Proving Fluid Tightness

Most valves are intended for duties for which steam-class fluid tightness is satisfactory. Tests for proving this degree of fluid tightness are normally

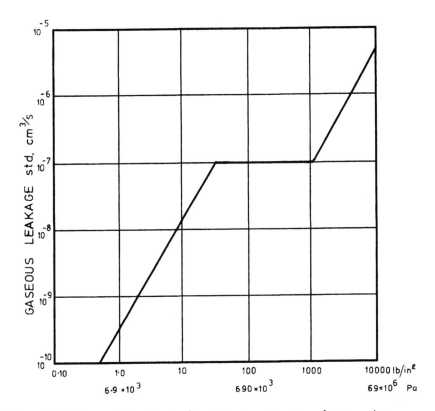

Figure 2-1. Proposed Zero Gas Leakage Criterion. (*Courtesy of Jet Propulsion Laboratory, California Institute of Technology. Reproduced from JPL Technical Report No. 32-926.*)

carried out with water, air, or inert gas. The tests are applied to the valve body and the seat, and depending on the construction of the valve, also to the stuffing box back seat, but they frequently exclude the stuffing box seal itself. When testing with water, the leakage rate is metered in terms of either volume-per-time unit or liquid droplets per time unit. Gas leakage may be metered by conducting the leakage gas through either water or a bubble-forming liquid leak-detector agent, and then counting the leakage gas bubbles per time unit. Using the bubble-forming leakage-detector agent permits metering very low leakage rates, down to 1×10^{-2} or 1×10^{-4} sccs (standard cubic centimeters per second), depending on the skill of the operator.[3]

Lower leakage rates in the atom class may be detected by using a search gas in conjunction with a search-gas detector.

Specifications for proving leakage tightness may be found in valve standards or in the separate standards listed in Appendix C. A description of leakage testing methods for the atom class may be found in BS 3636.

SEALING MECHANISM

Sealability Against Liquids

The sealability against liquids is determined by the surface tension and the viscosity of the liquid.

When the leakage capillary is filled with gas, surface tension can either draw the liquid into the capillary or repel the liquid, depending on the angle of contact formed by the liquid with the capillary wall. The value of the contact angle is a measure of the degree of wetting of the solid by the liquid and is indicated by the relative strength of the attractive forces exerted by the capillary wall on the liquid molecules, compared with the attractive forces between the liquid molecules themselves.

Figure 2-2 illustrates the forces acting on the liquid in the capillary. The opposing forces are in equilibrium if

$$\pi r^2 \Delta P = 2\pi r T \cos\theta \quad \text{or} \quad \Delta P = \frac{2T\cos\theta}{r} \qquad (2-1)$$

where
$\quad$ r = radius of capillary
$\quad \Delta P$ = capillary pressure
$\quad$ T = surface tension
$\quad \theta$ = contact angle between the solid and liquid

Thus, if the contact angle formed between the solid and liquid is greater than 90°, surface tension can prevent leakage flow. Conversely, if the contact angle is less than 90°, the liquid will draw into the capillaries and leakage flow will start at low pressures.

The tendency of metal surfaces to form a contact angle with the liquid of greater than 90° depends on the presence of a layer of oily, greasy, or waxy substances that normally cover metal surfaces. When this layer is removed by a solvent, the surface properties alter, and a liquid that previously was repelled may now wet the surface. For example, kerosene dissolves a greasy surface film, and a valve that originally was fluid-tight against water may

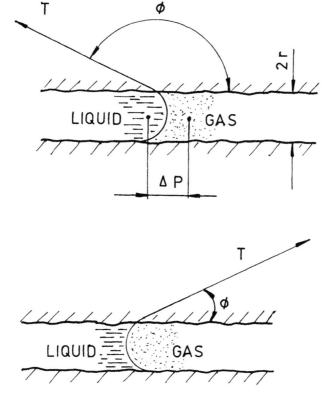

Figure 2-2. Effect of Surface Tension on Leakage Flow through Capillary.

leak badly after the seatings have been washed with kerosene. Wiping the seating surfaces with an ordinary cloth may be sufficient to restore the greasy film and, thus, the original seat tightness of the valve against water.

Once the leakage capillaries are flooded, the capillary pressure becomes zero, unless gas bubbles carried by the fluid break the liquid column. If the diameter of the leakage capillary is large, and the Reynolds number of the leakage flow is higher than critical, the leakage flow is turbulent. As the diameter of the capillary decreases and the Reynolds number decreases below its critical value, the leakage flow becomes laminar. This leakage flow will, from Poisuille's equation, vary inversely with the viscosity of the liquid and the length of the capillary and proportionally to the driving force and the diameter of the capillary. Thus, for conditions of high viscosity and small capillary size, the leakage flow can become so small that it reaches undetectable amounts.

Sealability Against Gases

The sealability against gases is determined by the viscosity of the gas and the size of the gas molecules. If the leakage capillary is large, the leakage flow will be turbulent. As the diameter of the capillary decreases and the Reynolds number decreases below its critical value, the leakage flow becomes laminar, and the leakage flow will, from Poisuille's equation, vary inversely with the viscosity of the gas and the length of the capillary, and proportionally to the driving force and the diameter of the capillary. As the diameter of the capillary decreases still further until it is of the same order of magnitude as the free mean path of the gas molecules, the flow loses its mass character and becomes diffusive, that is, the gas molecules flow through the capillaries by random thermal motion. The size of the capillary may decrease finally below the molecular size of the gas, but even then, flow will not strictly cease, since gases are known to be capable of diffusing through solid metal walls.

Mechanism for Closing Leakage Passages

Machined surfaces have two components making up their texture: a waviness with a comparatively wide distance between peaks, and a roughness consisting of very small irregularities superimposed on the wavy pattern. Even for the finest surface finish, these irregularities are large compared with the size of a molecule.

If the material of one of the mating bodies has a high enough yield strain, the leakage passages formed by the surface irregularities can be closed by elastic deformation alone. Rubber, which has a yield strain of approximately 1,000 times that of mild steel, provides a fluid-tight seal without being stressed above its elastic limit. Most materials, however, have a considerably lower elastic strain, so the material must be stressed above its elastic limit to close the leakage passages.

If both surfaces are metallic, only the summits of the surface irregularities meet initially, and small loads are sufficient to deform the summits plastically. As the area of real contact grows, the deformation of the surface irregularities becomes plastic-elastic. When the gaps formed by the surface waviness are closed, only the surface roughness in the valleys remains. To close these remaining channels, very high loads must be applied that may cause severe plastic deformation of the underlying material. However, the intimate contact between the two faces needs to extend only along a continuous line or ribbon to produce a fluid-tight seal. Radially directed asperities are difficult or impossible to seal.

VALVE SEATINGS

The effectiveness of the seating and the subsequent sealing of a valve are very important factors in the selection of a valve for a specific process function. Valve seatings are the portions of the seat and closure member that contact each other for closure. Because the seatings are subject to wear during the making of the seal, the sealability of the seatings tends to diminish with operation.

Metal Seatings

Operational wear is not limited to soft seated valves and it can be experienced with metal-seated valves if the process system is carrying a corrosive fluid or a fluid that contains particles. Metal seatings are prone to deformation by trapped fluids and wear particles. They are further damaged by corrosion, erosion, and abrasion. If the wear-particle size is large compared with the size of the surface irregularities, the surface finish will deteriorate as the seatings wear in. On the other hand, if the wear-particle size is small compared with the size of the surface irregularities, a coarse finish tends to improve as the seatings wear in. The wear-particle size depends not only on the type of the material and its condition, but also on the lubricity of the fluid and the contamination of the seatings with corrosion and fluid products, both of which reduce the wear-particle size.

The seating material must therefore be selected for resistance to erosion, corrosion, and abrasion. If the material fails in one of these requirements, it may be completely unsuitable for its duty. For example, corrosive action of the fluid greatly accelerates erosion. Similarly, a material that is highly resistant to erosion and corrosion may fail completely because of poor galling resistance. On the other hand, the best material may be too expensive for the class of valve being considered, and compromise may have to be made.

Table 2-1 gives data on the resistance of a variety of seating materials to erosion by jets of steam. Stainless steel AISI type 410 (13 Cr) in heat-treated form is shown to be particularly impervious to attack from

Table 2-1
Erosion Penetration
(Courtesy Crane Co.)

Resulting from the impingement of a 1.59 mm ($\frac{1}{16}$ inch) diameter jet of saturated steam of 2.41 MPa (350 psi) pressure for 100 hours on to a specimen 0.13 mm (0.005 inch) away from the orifice:

Class 1—less than 0.0127 mm (0.0005 inch) penetration
Stainless steel AISI tp 410 (13Cr) bar forged and heat treated
Delhi hard (17Cr)
Stainless steel AISI tp 304 (18Cr, 10Ni) cast
Stellite No. 6

Class 2—0.0127 mm (0.0005 inch) to 0.0254 mm (0.001 inch) penetration
Stainless steel AISI tp 304 (18Cr, 10Ni) wrought
Stainless steel AISI tp 316 (18Cr, 12Ni, 2.4Mo) arc deposit
Stellite No. 6 torch deposit

Class 3—0.0254 mm (0.001 inch) to 0.0508 mm (0.002 inch) penetration
Stainless steel AISI tp 410 (13Cr) forged, hardened 444 Bhn
Nickel—base copper—tin alloy
Chromium plate on No. 4 brass (0.0254 mm = 0.001 inch)

Class 4—0.0508 mm (0.002 inch) to 0.1016 mm (0.004 inch) penetration
Brass stem stock
Nitralloy $2\frac{1}{2}$ Ni
Nitalloy high carbon and chrome
Nitralloy Cr—V sorbite—ferrite lake structure, annealed after nitriding 950 Bhn
Nitralloy Cr—V Bhn 770 sorbitic structure
Nitralloy Cr—Al Bhn 758 ferritic structure
Monel modifications

Class 5—0.1016 mm (0.004 inch) to 0.2032 mm (0.008 inch) penetration
Brass No. 4, No. 5, No. 22, No. 24
Nitralloy Cr—Al Bhn 1155 sorbitic structure
Nitralloy Cr—V Bhn 739 ferrite lake structure
Monel metal, cast

Class 6—0.2032 mm (0.008 inch) to 0.4064 mm (0.016 inch) penetration
Low alloy steel C 0.16, Mo 0.27, Si 0.19, Mn 0.96
Low alloy steel Cu 0.64, Si 1.37, Mn 1.42
Ferro steel

Class 7—0.4064 mm (0.016 inch) to 0.8128 mm (0.032 inch) penetration
Rolled red brass
Grey cast iron
Malleable iron
Carbon steel 0.40 C

steam erosion. However, if the fluid lacks lubricity, type 410 stainless in like contact offers only fair resistance to galling unless the mating components are of different hardness. For steam and other fluids that lack lubricity, a combination of type 410 stainless steel and copper-nickel alloy is frequently used. Stellite, a cobalt-nickel-chromium alloy, has proved most successful against erosion and galling at elevated temperatures, and against corrosion for a wide range of corrosives.

API Std 600 lists seating materials and their combinations frequently used in steel valves.

Sealing with Sealants

Certain valves have the facility to periodically introduce sealants into the valve seat and stems to maintain an effective seal over an extended period. The leakage passages between metal seatings can be closed by sealants injected into the space between the seatings after the valve has been closed. One metal-seated valve that relies completely on this sealing method is the lubricated plug valve. The injection of a sealant to the seatings is used also in some other types of valves to provide an emergency seat seal after the original seat seal has failed.

Soft Seatings

Soft seats are very effective, but they have limited use at high temperatures and pressures. Manufacturers of proprietary soft seats will state the maximum and minimum design pressures and temperatures for which their products are suitable. Some soft seats are also not suitable for some fluids at certain pressures and temperatures.

In the case of soft seatings, one or both seating faces may consist of a soft material such as plastic or rubber. Because these materials conform readily to the mating face, soft seated valves can achieve an extremely high degree of fluid tightness. Also, the high degree of fluid tightness can be achieved repeatedly. On the debit side, the application of these materials is limited by their degree of compatibility with the fluid and by temperature.

A sometimes unexpected limitation of soft seating materials exists in situations in which the valve shuts off a system that is suddenly filled with gas at high pressure. The high-pressure gas entering the closed system acts like a piston on the gas that filled the system. The heat of compression can be high enough to disintegrate the soft seating material.

Table 2-2 indicates the magnitude of the temperature rise that can occur. This particular list gives the experimentally determined temperature rise of oxygen that has been suddenly pressurized from an initial state of atmospheric pressure and 15°C.[4]

Heat damage to the soft seating element is combated in globe valves by a heat sink resembling a metallic button with a large heat-absorbing surface, which is located ahead of the soft seating element. In the case of oxygen service, this design measure may not be enough to prevent the soft seating element from bursting into flames. To prevent such failure, the valve inlet passage may have to be extended beyond the seat passage, so that the end of the inlet passage forms a pocket in which the high temperature gas can accumulate away from the seatings.

In designing soft seatings, the main consideration is to prevent the soft seating element from being displaced or extruded by the fluid pressure.

GASKETS

Flat Metallic Gaskets

Flat metallic gaskets adapt to the irregularities of the flange face by elastic and plastic deformation. To inhibit plastic deformation of the flange face, the yield shear strength of the gasket material must be considerably lower than that of the flange material.

Table 2-2
Experimentally Determined Temperature Rise of Oxygen Due to Sudden Pressurizing from an Initial State of Atmospheric Pressure and 15°C

Sudden Pressure Rise		Temperature Rise	
25	Bar (360 lb/in^2)	375°C	(705°F)
50	Bar (725 lb/in^2)	490°C	(915°F)
100	Bar (1450 lb/in^2)	630°C	(1165°F)
150	Bar (2175 lb/in^2)	730°C	(1345°F)
200	Bar (2900 lb/in^2)	790°C	(1455°F)

The free lateral expansion of the gasket due to yielding is resisted by the roughness of the flange face. This resistance to lateral expansion causes the yield zone to enter the gasket from its lateral boundaries, while the remainder of the gasket deforms elastically initially. If the flange face is rough enough to prevent slippage of the gasket altogether—in which case the friction factor is 0.5—the gasket will not expand until the yield zones have met in the center of the gasket.[5]

For gaskets of a non-strain-hardening material mounted between perfectly rough flange faces, the mean gasket pressure is, according to Lok,[1] approximately:

$$P_m = 2k \left(1 + \frac{w}{4t}\right) \qquad (2-2)$$

where
P_m = mean gasket pressure
k = yield shear stress of gasket material
w = gasket width
t = gasket thickness

If the friction factor were zero, the gasket pressure could not exceed twice the yield shear stress. Thus, a high friction factor improves the load-bearing capacity of the gasket.

Lok has also shown that a friction factor lower than 0.5, but not less than 0.2, diminishes the load-bearing capacity of the gasket only by a small amount. Fortunately, the friction factor of finely machined flange faces is higher than 0.2. But the friction factor for normal aluminum gaskets in contact with lapped flange faces has been found to be only 0.05. The degree to which surface irregularities are filled in this case is very low. Polishing the flange face, as is sometimes done for important joints, is therefore not recommended.

Lok considers spiral grooves with an apex angle of 90° and a depth of 0.1 mm (125 grooves per inch) representative for flange face finishes in the steam class, and a depth of 0.01 mm (1250 grooves per inch) representative in the atom class. To achieve the desired degree of filling of these grooves, Lok proposes the following dimensional and pressure-stress relationships.

for steam class: $\quad \dfrac{w}{t} > 5 \quad$ and $\quad \dfrac{P_m}{2k} > 2.25$

for atom class: $\quad \dfrac{w}{t} > 16 \quad$ and $\quad \dfrac{P_m}{2k} > 5$

Gaskets of Exfoliated Graphite[6]

Exfoliated graphite is manufactured by the thermal exfoliation of graphite intercalation compounds and then calendered into flexible foil and laminated without an additional binder. The material thus produced possesses extraordinary physical and chemical properties that render it particularly suitable for gaskets. Some of theses properties are:

* High impermeability to gases and liquids, irrespective of temperature and time.
* Resistance to extremes of temperature, ranging from −200°C (−330°F) to 500°C (930°F) in oxidizing atmosphere and up to 3,000°C (5,430°F) in reducing or inert atmosphere.
* High resistance to most reagents, for example, inorganic or organic acids and bases, solvents, and hot oils and waxes. (Exceptions are strongly oxidizing compounds such as concentrated nitric acid, highly concentrated sulfuric acid, chromium (VI)-permanganate solutions, chloric acid, and molten alkaline and alkaline earth metals).
* Graphite gaskets with an initial density of 1.0 will conform readily to irregularities of flange faces, even at relatively low surface pressures. As the gasket is compressed further during assembly, the resilience increases sharply, with the result that the seal behaves dynamically. This behavior remains constant from the lowest temperature to more than 3,000°C (5,430°F). Thus graphite gaskets absorb pressure and temperature load changes, as well as vibrations occurring in the flange.
* The ability of graphite gaskets to conform relatively easily to surface irregularities makes these gaskets particularly suitable for sensitive flanges such as enamel, glass, and graphite flanges.
* Large gaskets and those of complicated shape can be constructed simply from combined segments that overlap. The lapped joints do not constitute weak points.
* Graphite can be used without misgivings in the food industry.

Common gasket constructions include:

* Plain graphite gaskets
* Graphite gaskets with steel sheet inserts
* Graphite gaskets with steel sheet inserts and inner or inner and outer edge cladding

- Grooved metal gaskets with graphite facings
- Spiral wound gaskets

Because of the graphite structure, plain graphite gaskets are sensitive to breakage and surface damage. For this reason, graphite gaskets with steel inserts and spiral wound gaskets are commonly preferred. There are, however, applications where the unrestrained flexibility of the plain graphite gasket facilitates sealing.

Spiral Wound Gaskets

Spiral wound gaskets consist of a V-shaped metal strip that is spirally wound on edge, and a soft filler inlay between the laminations. Several turns of the metal strip at start and finish are spot welded to prevent the gasket from unwinding. The metal strip provides a degree of resiliency to the gasket, which compensates for minor flange movements; whereas, the filler material is the sealing medium that flows into the imperfections of the flange face.

Manufacturers specify the amount of compression for the installed gasket to ensure that the gasket is correctly stressed and exhibits the desired resiliency. The resultant gasket operating thickness must be controlled by controlled bolt loading, or the depth of a recess for the gasket in the flange, or by inner and/or outer compression rings. The inner compression ring has the additional duty of protecting the gasket from erosion by the fluid, while the outer compression ring locates the gasket within the bolt diameter.

The load-carrying capacity of the gasket at the operating thickness is controlled by the number of strip windings per unit width, referred to as gasket density. Thus, spiral wound gaskets are tailor-made for the pressure range for which they are intended.

The diametrical clearance for unconfined spiral wound gaskets between pipe bore and inner gasket diameter, and between outer gasket diameter and diameter of the raised flange face, should be at least 6 mm ($\frac{1}{4}$ in). If the gasket is wrongly installed and protrudes into the pipe bore or over the raised flange face, the sealing action of the gasket is severely impaired. The diametrical clearance recommended for confined gaskets is 1.5 mm ($\frac{1}{16}$ in).

The metal windings are commonly made of stainless steel or nickel-based alloys, which are the inventory materials of most manufacturers. The windings may be made also of special materials such as mild steel, copper, or even gold or platinum. In selecting materials for corrosive fluids or high temperatures, the resistance of the material to stress corrosion or intergranular corrosion must be considered. Manufacturers might be able to advise on the selection of the material.

The gasket filler material must be selected for fluid compatibility and temperature resistance. Typical filler materials are PTFE (polytetrafluoroethylene), pure graphite, mica with rubber or graphite binder, and ceramic fiber paper. Manufacturers will advise on the field of application of each filler material.

The filler material also affects the sealability of the gasket. Gaskets with asbestos and ceramic paper filler materials require higher seating stresses than gaskets with softer and more impervious filler materials to achieve comparable fluid tightness. They also need more care in the selection of the flange surface finish.

In most practical applications, the user must be content with flange face finishes that are commercially available. For otherwise identical geometry of the flange-sealing surface, however, the surface roughness may vary widely, typically between 3.2 and 12.5 μm Ra (125 and 500 μin. Ra). Optimum sealing has been achieved with a finish described in ANSI B16.5, with the resultant surface finish limited to the 3.2 to 6.3 μm Ra (125 to 250 μin. Ra) range. Surface roughness higher than 6.3 μm Ra (250 μin. Ra) may require unusually high seating stresses to produce the desired flange seal. On the other hand, surface finishes significantly smoother than 3.2 μm Ra (125 μin. Ra) may result in poor sealing performance, probably because of insufficient friction between gasket and flange faces to prevent lateral displacement of the gasket.

A manufacturer's publication dealing with design criteria of spiral wound gaskets may be found in Reference 7.

Gasket Blowout

Unconfined gaskets in flanged joints may blow out prior to leakage warning when inadequately designed.

This mode of gasket failure will not occur if the friction force at the gasket faces exceeds the fluid force acting on the gasket in the radial direction, as expressed by the equation:

$$2\mu F \geq Pt\pi d_m \quad \text{or} \quad F \geq \frac{Pt\pi d_m}{2\mu} \tag{2-3}$$

where

   μ = friction factor
   F = gasket working load
   P = fluid gauge pressure
   t = gasket thickness
   d_m = mean gasket diameter

The joint begins to leak if:

$$F \geq wPm\pi d_m \qquad (2-4)$$

in which

   m = gasket factor
   w = gasket width

   The gasket factor is a measure of the sealing ability of the gasket, and defines the ratio of residual gasket stress to the fluid pressure at which leakage begins to develop. Its value is found experimentally.

   It follows thus from Equations 2-3 and 2-4 that the gasket is safe against blowout without prior leakage warning if:

$$w \geq \frac{t}{2\mu m} \qquad (2-5)$$

   Krägeloh[8] regarded a gasket factor of 1.0 and a friction factor of 0.1 safe for most practical applications. Based on these factors, the width of the gasket should be not less than five times its thickness to prevent blowout of the gasket without prior leakage warning.

VALVE STEM SEALS

Compression Packings

Construction. Compression packings are the sealing elements in stuffing boxes (see Figures 3-17 through 3-19). They consist of a soft material that is stuffed into the stuffing box and compressed by a gland to form a seal around the valve stem.

   The packings may have to withstand extremes of temperature, be resistant to aggressive media, display a low friction factor and adequate structural strength, and be impervious to the fluid to be sealed. To meet this wide range of requirements, and at the same time offer economy of use, innumerable types of packing constructions have evolved.

The types of lubricants used for this purpose are oils and greases when water and aqueous solutions are to be sealed, and soaps and insoluble substances when fluids like oil or gasoline are to be sealed. Unfortunately, liquid lubricants tend to migrate under pressure, particularly at higher temperatures, causing the packing to shrink and harden. Such packings must, therefore, be retightened from time to time to make up for loss of packing volume. To keep this loss to a minimum, the liquid content of valve stem packings is normally held to 10% of the weight of the packing.

With the advent of PTFE, a solid lubricant became available that can be used in fibrous packings without the addition of a liquid lubricant.

Asbestos is now avoided in packings where possible, replaced by polymer filament yarns, such as PTFE and aramid, and by pure graphite fiber or foil. Other packing materials include vegetable fibers such as cotton, flax, and ramie (frequently lubricated with PTFE), and twisted and folded metal ribbons.

The types of fibrous packing constructions in order of mechanical strength are loose fill, twisted yarn, braid over twisted core, square-plait braid, and interbraid constructions. The covers of the latter three types of packing constructions often contain metal wire within the strands to increase the mechanical strength of the packing for high fluid pressure and high temperature applications.

Reference 9 offers advice on selection and application of compression packings. Standards on packings may be found in Appendix C.

Sealing action. The sealing action of compression packings is due to their ability to expand laterally against the stem and stuffing box walls when stressed by tightening of the gland.

The stress exerted on the lateral faces of a confined elastic solid by an applied axial stress depends on Poisson's ratio for the material, as expressed by:

$$\sigma_1 = \sigma_a \left(\frac{1 - \mu}{\mu} \right) \qquad\qquad (2\text{--}6)$$

where
σ_1 = lateral stress
σ_a = axial stress
μ = Poisson's ratio

= ratio of lateral expansion to axial compression of an elastic solid compressed between two faces

Thus, the lateral stress equals the axial stress only if $\mu = 0.5$, in which case the material is incompressible in bulk.

A material with a Poisson's ratio nearly equal to 0.5 is soft rubber, and it is known that soft rubber transmits pressure in much the same way as a liquid.[10] Solid PTFE has Poisson's ratio of 0.46 at 23°C (73°F) and 0.36 at 100°C (212°F).[11] A solid PTFE packing is capable of transmitting 85% and 56% of the axial stress to the lateral faces at the respective temperatures. Other packing materials, however, are much more compressible in bulk, so Poisson's ratio, if it can be defined for these materials, is considerably less than 0.5.

When such packing is compressed in the stuffing box, axial shrinkage of the packing causes friction between itself and the side walls that prevents the transmission of the full gland force to the bottom of the packing. This fall in axial packing pressure is quite rapid, and its theoretical value can be calculated.[12,13]

The theoretical pressure distribution, however, applies to static conditions only. When the stem is being moved, a pressure distribution takes place so that an analysis of the actual pressure distribution is difficult.

The pressure distribution is also influenced by the mode of packing installation. If the packing consists of a square cord, bending of the packing around the stem causes the packing to initially assume the shape of a trapezoid. When compressing the packing, the pressure on the inner periphery will be higher than on the outer periphery.

When the fluid pressure applied to the bottom of the packing begins to exceed the lateral packing pressure, a gap develops between the packing and the lateral faces, allowing the fluid to enter this space. In the case of low-pressure applications, the gland may finally have to be retightened to maintain a fluid seal.

When the fluid pressure is high enough, the sealing action takes place just below the gland, where the fluid pressure attempts to extrude the packing through the gland clearances. At this stage, the sealing action has become automatic.

Readings of the fluid pressure gradient of leakage flow along the stuffing box of rotating shafts, as shown in Figure 2-3, confirm this function of the stuffing box seal.[12,13] The pressure gradient at low fluid pressures is more or less uniform, which indicates little influence by the fluid pressure on the sealing action. On the other hand, the readings at high fluid pressure

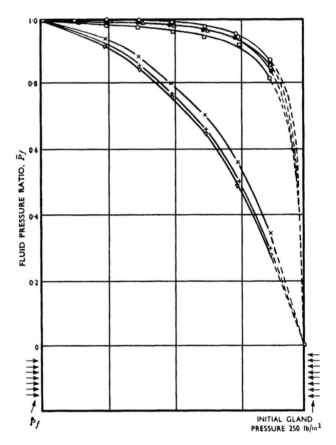

Figure 2-3. Distribution of Fluid Pressure for Four Rings of PTFE-Impregnated Plaited Cotton Packing Where $\hat{P}_f$ = Applied Fluid Pressure and $\bar{P}_f$ = Normalized Fluid Pressure $\hat{P}_f/\hat{P}_f$ and P_f = Fluid Pressure. Each Set of Measurements Taken 6 Hours After Change of Pressure. Shaft Speed: 850 rev/min. Applied Gland Pressure: 250 lb/in². Water pressure, lb/in²: ○ 1000, △ 700, ● 400, □ 250, × 75, + 26,2. (*Reprinted from Proceedings of the Institution of Mechanical Engineers, London, 174 No. 6, 1960, p. 278, by D. F. Denny; and D. E. Tumbull.*)

show that 90% of the pressure drop occurs across the packing ring just below the gland. This indicates a dominant influence of the fluid pressure on the sealing action.

In the case of high fluid pressures, therefore, the packing ring just below the gland is the most important one, and must be selected for resistance to extrusion and wear and be carefully installed. Also, extra long stuffing boxes for high-pressure applications do not serve the intended purpose.

If the packing is incompressible in bulk, as in the case of soft rubber, the axial packing pressure introduced by tightening of the gland will produce a uniform lateral packing pressure over the entire length of the packing. Fluid pressure applied to the bottom of the packing increases the lateral packing pressure by the amount of fluid pressure, so the sealing action is automatic once interference between packing and the lateral restraining faces has been established.

Unfortunately, rubber tends to grip the stem and impede its operation unless the inner face of the rubber packing is provided with a slippery surface. For this reason, rubber packings are normally used in the form of O-rings, which because of their size offer only a narrow contact face to the stem.

Corrosion of stainless steel valve stems by packings. Stainless steel valve stems—in particular those made of AISI type 410 (13Cr) steel—corrode frequently where the face contacts the packing. The corrosion occurs usually during storage preceding service, when the packing is saturated with water from the hydrostatic test.

If the valve is placed into service immediately after the hydrostatic test, no corrosion occurs.[14] H. J. Reynolds, Jr. has published the results of his investigations into this corrosion phenomenon; the following is an abstract.[15] Corrosion of stainless steel valve stems underlying wet packing is theorized to be the result of the deaerated environment imposed on the steel surface by the restricting packing—an environment that influences the active-passive nature of the metal. Numerous small anodes are created at oxygen-deficient sensitive points of the protective oxide surface film on the stainless steel. These, along with large masses of retained passive metal acting as cathodes, result in galvanic cell action within the metal. Graphite, often contained in the packing, acts as a cathodic material to the active anodic sites on the steel, and appreciably aggravates the attack at the initial corrosion sites through increased galvanic current density.

Because of the corrosion mechanism involved, it is impractical to make an effective non-corrosive packing using so-called non-corrosive ingredients. Incorporating a corrosion inhibitor into the packing is thus required, which will influence the anodic or cathodic reactions to produce a minimum corrosion rate. Of the anodic inhibitors evaluated, only those containing an oxidizing anion, such as sodium nitrite, are efficient. Cathodic protection by sacrificial metals such as zinc, contained in the packing, also provides good corrosion control. Better protection with a minimum effect on compression and serviceability characteristics of the packing is provided by homogeneously dispersed sodium nitrite and a zinc-dust interlayer incorporated into the material.

High chromium-content stainless steels—especially those containing nickel—exhibit a marked increase in resistance to corrosion by inhibited packing, presumably because of the more rapidly protective oxide surface film and better retention of the passivating film.

Lip-Type Packings

Lip-type packings expand laterally because of the flexibility of their lips, which are forced against the restraining side walls by the fluid pressure. This mode of expansion of the packing permits the use of relatively rigid construction materials, which would not perform as well in compression packings. On the debit side, the sealing action of lip-type packings is in one direction only.

Most lip-type packings for valve stems are made of virgin or filled PTFE. However, fabric-reinforced rubber and leather are also used, mainly for hydraulic applications. Most lip-type packings for valve stems are V shaped, because they accommodate themselves conveniently in narrow packing spaces.

The rings of V-packings made of PTFE and reinforced rubber are designed to touch each other on small areas near the tips of their lips, and large areas are separated by a gap that permits the fluid pressure to act freely on the lips. Leather V-packing rings lack the rigidity of those made of PTFE and reinforced rubber, and are therefore designed to fully support each other.

V-packings made of PTFE and reinforced rubber are commonly provided with flared lips that automatically preload the restraining lateral faces. In this case, only slight initial tightening of the packing is necessary to achieve a fluid seal. V-packing rings made of leather have straight walls and require a slightly higher axial preload. If a low packing friction is important, as in automatic control valves, the packing is frequently loaded from the bottom by a spring of predetermined strength to prevent manual overloading of the packing.

Squeeze-Type Packings

The name squeeze-type packing applies to O-ring packings and the like. Such packings are installed with lateral squeeze, and rely on the elastic strain of the packing material for the maintenance of the lateral preload. When the fluid pressure enters the packing housing from the bottom, the packing moves towards the gap between the valve stem and the back-up support and thereby plugs the leakage path. When the packing housing is depressurized again, the packing regains its original configuration. Because elastomers display the high-yield strain necessary for this mode of action, most squeeze packings are made of these materials.

Extrusion of the packing is controlled by the width of the clearance gap between the stem and the packing back-up support, and by the rigidity of the elastomer as expressed by the modulus of elasticity. Manufacturers express the rigidity of elastomers conventionally in terms of Durometer hardness, although Durometer hardness may express different moduli of elasticity for different classes of compounds. Very small clearance gaps are controlled by leather or plastic back-up rings, which fit tightly around the valve stem. Manufacturers of O-ring packings supply tables, which relate the Durometer hardness and the clearance gap around the stem to the fluid pressure at which the packing is safe against extrusion.

Thrust Packings

Thrust packings consist of a packing ring or washer mounted between shoulders provided on bonnet and valve stem, whereby the valve stem is free to move in an axial direction against the packing ring. The initial stem seal may be provided either by a supplementary radial packing such as a compression packing, or by a spring that forces the shoulder of the stem against the thrust packing. The fluid pressure then forces the shoulder of the stem into more intimate contact with the packing.

Thrust packings are found frequently in ball valves such as those shown in Figures 3-61 through 3-63, 3-65, and 3-67.

Diaphragm Valve Stem Seals

Diaphragm valve stem seals represent flexible pressure-containing valve covers, which link the valve stem with the closure member. Such seals prevent any leakage past the stem to the atmosphere, except in the case of a fracture of the diaphragm. The shape of the diaphragm may represent a dome, as in the valve shown in Figure 3-7,[16] or a bellows, as in the valves shown in Figures 3-6 and 3-39. Depending on the application of the valve, the construction material of the diaphragm may be stainless steel, a plastic, or an elastomer.

Dome-shaped diaphragms offer a large uncompensated area to the fluid pressure, so the valve stem has to overcome a correspondingly high fluid load. This restricts the use of dome-shaped diaphragms to smaller valves, depending on the fluid pressure. Also, because the possible deflection of dome-shaped diaphragms is limited, such diaphragms are suitable only for short lift valves.

Bellows-shaped diaphragms, on the other hand, offer only a small uncompensated area to the fluid pressure, and therefore transmit a correspondingly lower fluid load to the valve stem. This permits bellows-shaped diaphragms to be used in larger valves. In addition, bellows-shaped diaphragms may be adapted to any valve lift.

To prevent any gross leakage to the atmosphere from a fracture of the diaphragm, valves with diaphragm valve stem seals are frequently provided with a secondary valve stem seal such as a compression packing.

FLOW THROUGH VALVES

Valves may be regarded as analogous to control orifices in which the area of opening is readily adjustable. As such, the friction loss across the valve varies with flow, as expressed by the general relationship

$$v \propto (\Delta h)^{1/2}$$

$$v \propto (\Delta p)^{1/2}$$

where

    v = flow velocity
    Δh = headloss
    Δp = pressure loss

For any valve position, numerous relationships between flow and flow resistance have been established, using experimentally determined resistance or flow parameters. Common parameters so determined are the resistance coefficient ζ and, dependent on the system of units, the flow coefficients C_v, K_v, and A_v. It is standard practice to base these parameters on the nominal valve size.[17,18]

Resistance Coefficient ζ

The resistance coefficient ζ defines the friction loss attributable to a valve in a pipeline in terms of velocity head or velocity pressure, as expressed by the equations

$$\Delta h = \zeta \frac{v^2}{2g} \text{ (coherent SI or imperial units)} \tag{2-7}$$

$$\text{and: } \Delta p = \zeta \frac{v^2 \rho}{2} \text{ (coherent SI units)} \tag{2-8}$$

$$\text{or: } \Delta p = \zeta \frac{v^2 \rho}{2g} \text{ (coherent imperial units)} \tag{2-9}$$

where
ρ = density of fluid
g = local acceleration due to gravity

The equations are valid for single-phase flow of Newtonian liquids and for both turbulent and laminar flow conditions. They may also be used for flow of gas at low Mach numbers. As the Mach number at the valve inlet approaches 0.2, the effects of compressibility become noticeable but are unlikely to be significant even for Mach numbers up to 0.5.[17]

Valves of the same type but of different manufacture, and also of the same line but different size, are not normally geometrically similar. For this reason, the resistance coefficient of a particular size and type of valve can differ considerably between makes. Table 2-3 can therefore provide only typical resistance-coefficient values. The values apply to fully open valves only and for Re $\geq 10^4$. Correction factor K_1 for partial valve opening may be obtained from Figure 2-4 through 2-7.

The Engineering Sciences Data Unit, London,[17] deals more comprehensively with pressure losses in valves. Their publication also covers correction factors for Re $< 10^4$ and shows the influence of valve size on the ζ-Value and scatter of data, as obtained from both published and unpublished reports and from results obtained from various manufacturers.

In the case of partially open valves and valves with reduced seat area, as in valves with a converging/diverging flow passage, the energy of the flow stream at the vena contracta converts partially back into static energy.

Table 2-3
**Approximate Resistance Coefficients of Fully Open Valves Under
Conditions of Fully Turbulent Flow**

Globe valve, standard pattern:	
• Full bore seat, cast.	$\zeta = 4.0–10.0$
• Full bore seat, forged (small sizes only).	$\zeta = 5.0–13.0$
Globe valve, 45° oblique pattern:	
• Full bore seat, cast.	$\zeta = 1.0–3.0$
Globe valve, angle pattern:	
• Full bore seat, cast.	$\zeta = 2.0–5.0$
• Full bore seat, forged (small sizes only).	$\zeta = 1.5–3.0$
Gate valve, full bore:	$\zeta = 0.1–0.3$
Ball valve, full bore:	$\zeta = 0.1$
Plug valve, rectangular port:	
• Full flow area.	$\zeta = 0.3–0.5$
• 80% flow area.	$\zeta = 0.7–1.2$
• 60% flow area.	$\zeta = 0.7–2.0$
Plug valve, circular port, full bore:	$\zeta = 0.2–0.3$
Butterfly valve, dependent on blade thickness:	$\zeta = 0.2–1.5$
Diaphragm valve:	
• Weir type.	$\zeta = 2.0–3.5$
• Straight-through type.	$\zeta = 0.6–0.9$
Lift check valve (as globe valve):	
Swing check valve:	$\zeta = 1.0$
Tilting-disc check valve:	$\zeta = 1.0$

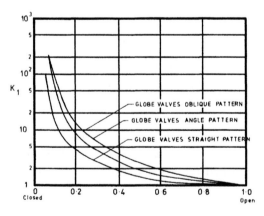

Figure 2-4. Approximate Effect of Partial Opening of Globe Valves on Resistance Coefficient. (*Courtesy of Engineering Sciences Data Unit. Reproduced from Item No. 69022, Figure 13.*)

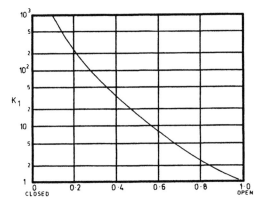

Figure 2-5. Approximate Effect of Partial Opening of Gate Valves on Resistance Coefficient. (*Courtesy of Engineering Sciences Data Unit. Reproduced from Item No. 69022, Figure 14.*)

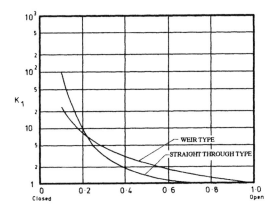

Figure 2-6. Approximate Effect of Partial Opening of Diaphragm Valves on Resistance Coefficient. (*Courtesy of Engineering Sciences Data Unit. Reproduced from Item No. 69022, Figure 15.*)

Figure 2-8 shows the influence of the pressure recovery on the resistance coefficient of fully open venturi-type gate valves in which the gap between the seats is bridged by an eyepiece.

The amount of static energy recovered depends on the ratio of the diameters of the flow passage (d^2/D^2), the taper angle ($\alpha/2$) of the diffuser, and the length (L) of straight pipe after the valve throat in terms of pipe diameter (d = valve throat diameters, and D = pipe diameter).

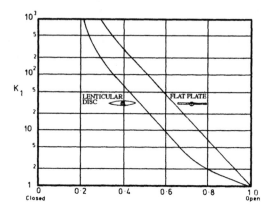

Figure 2-7. Approximate Effect of Partial Opening of Butterfly Valves on Resistance Coefficient. (*Courtesy of Engineering Sciences Data Unit. Reproduced from Item No. 69022, Figure 12.*)

If the straight length of pipe after the valve throat ≥ 12 D, the pressure loss cannot exceed the Borda-Carnot loss, which is:

$$\Delta h = \frac{(v_d - v_D)^2}{2g} \qquad (2-10)$$

where

v_d = flow velocity in valve throat

v_D = flow velocity in pipeline

This maximum pressure loss occurs if the taper angle ($\alpha/2$) of the diffuser > 30°. With a decreasing taper angle, the pressure loss decreases and reaches its lowest value at a taper angle of 4°.

If the straight pipe after the valve throat < 12 D, the pressure loss increases and reaches its maximum value when the static energy converted into kinetic energy gets completely lost, in which case:

$$\Delta h = \frac{(V_d^2 - V_D^2)}{2g} \qquad (2-11)$$

The friction loss approaches this maximum value if the valve with a converging/diverging flow passage is mounted directly against a header.

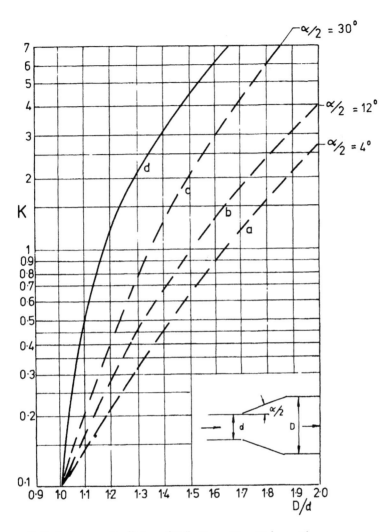

Figure 2-8. Resistance Coefficient of Fully Open Gate Valves with Converging-Diverging Flow Passage and Eye Piece Between the Seats. Curves a to c Apply to L ≥ 12 D, and Curve d for L = Zero, Where L = Straight Length of Pipe Downstream of Venturi Throat. (*Courtesy of VDI Verlag GmbH, Reproduced from BWK Arbeitsblatt 42, Dec. 1953, by H. Haferkamp and A. Kreuz.*)

Flow Coefficient C_v

The flow coefficient C_v states the flow capacity of a valve in gal (U.S.)/min of water at a temperature of 60°F that will flow through a valve with a pressure loss of one pound per square inch at a specific opening position, as defined by the equation:

$$C_v = Q \left(\frac{\Delta p_0}{\Delta p} \times \frac{\rho}{\rho_0} \right)^{1/2} \tag{2-12}$$

Where

Q = U.S. gal/min
Δp_0 = reference differential pressure = 1 lb/in.^{2}
Δp = operating differential pressure in lb/in.^{2}
ρ_0 = density of reference fluid, water = 62.4 lb/ft^3
ρ = density of operating fluid in lb/ft^3

Because ρ/ρ_0 = specific gravity and the numerical value of Δp_0 is unity, Equation 2-12 is normally presented in the form:

$$C_v = Q \left(\frac{G}{\Delta p} \right)^{1/2} \tag{2-13}$$

Where

G = specific gravity

This relationship is incorporated in the International Standards Association (ISA) standards S.39.1 to S.39.4 and Publication 534-1[19] of the International Electrotechnical Commission (IEC), and applies to single-phase and fully turbulent flow of Newtonian liquids. For other flow conditions, refer to the above ISA standards, IEC Publication 534-2 Parts 1 and 2, or Reference 20.

Flow Coefficient K_v

The flow coefficient K_v is a version of coefficient C_v in mixed SI units. It states the number of cubic meters per hour of water at a temperature between 5° and 40°C that will flow through the valve with a pressure loss of one bar at a specific opening position, as defined by the equation:

$$K_v = Q \left(\frac{\Delta p_0}{\Delta p} \times \frac{\rho}{\rho_0} \right)^{1/2} \tag{2-14}$$

where

Q = m³/hour
Δp_0 = reference differential pressure = 1 bar
Δp = operating differential pressure, bar
ρ_0 = density of reference fluid (water = 1,000 kg/m³)
ρ = density of operating fluid, kg/m³

Because ρ/ρ_0 = specific gravity and the numerical value of Δp_0 is unity, Equation 2-14 is normally presented in the form:

$$K_v = Q \left(\frac{G}{\Delta p} \right)^{1/2} \tag{2-15}$$

where

G = specific gravity

This relationship is contained in the IEC Publication 534-1[19] and applies to single-phase and fully turbulent flow of Newtonian liquids. For other flow conditions, refer to IEC Publication 534-2, Parts 1 and 2, or to manufacturer's catalogs.

Flow Coefficient A_v

The flow coefficient A_v is a version of the flow coefficient K_v in coherent SI units. A_v states the number of cubic meters per second of water at a temperature between 5° and 40°C that will flow through the valve with a pressure loss of one Pascal at a specific opening position, as defined by the equation:

$$A_v = Q \left(\frac{\rho}{\Delta p} \right)^{1/2} \tag{2-16}$$

where

Q = flow rate, m³/s
p = operational differential pressure, Pa
ρ = density of Newtonian liquid, kg/m³

A_v is derived from Equation 2-8, which may be presented in the following forms:

$$Q = A \left(\frac{2}{\zeta}\right)^{1/2} \left(\frac{\Delta p}{\rho}\right)^{1/2} \tag{2-17}$$

$$A \left(\frac{2}{\zeta}\right)^{1/2} = Q \left(\frac{\rho}{\Delta p}\right)^{1/2} \tag{2-18}$$

where
  v = flow velocity of fluid, m/s
  A = cross-sectional area, m^2

The expression $A \left(\dfrac{2}{\zeta}\right)^{1/2}$ is replaced in Equation 2-16 by a single expression A_v.

This relationship is contained in IEC Publication 534-1[19] and applies to single-phase and fully turbulent flow of Newtonian liquids. For other flow condition, see IEC Publication 534-2, Parts 1 and 2.

Interrelationships Between Resistance and Flow Coefficients

Resistance and flow coefficients are interrelated. If one coefficient is known, the other coefficients can be calculated. These are the interrelationships in which:

  d_{inch} = reference pipe bore, inches
  d_{mm} = reference pipe bore, mm

$$\zeta = \frac{889 d_{inch}^4}{C_v^2} = \frac{2.14 d_{mm}^4}{10^3 C_v^2} \qquad C_v = \frac{29.8 d_{inch}^2}{\zeta^{1/2}} = \frac{4.62 d_{mm}^2}{10^2 \zeta^{1/2}} \tag{2-19}$$

$$\zeta = \frac{665.2 d_{inch}^4}{K_v^2} = \frac{1.6 d_{mm}^4}{10^3 K_v^2} \qquad K_v = \frac{25.8 d_{inch}^2}{\zeta^{1/2}} = \frac{39.98 d_{mm}^4}{10^3 \zeta^{1/2}} \tag{2-20}$$

$$\zeta = \frac{512.\, d_{inch}^4}{10^9 A_v^2} = \frac{1.23 d_{mm}^4}{10^{12} A_v^2} \qquad A_v = \frac{716 d_{inch}^2}{10^6 \zeta^{1/2}} = \frac{1.11 d_{mm}^2}{10^6 \zeta^{1/2}} \tag{2-21}$$

$$\frac{K_v}{C_v} = 865 \times 10^{-3} \tag{2-22}$$

$$\frac{A_v}{C_v} = 23.8 \times 10^{-6} \tag{2-23}$$

$$\frac{A_v}{K_v} = 27.8 \times 10^{-6} \tag{2-24}$$

Relationship Between Resistance Coefficient and Valve Opening Position

The relationship between fractional valve opening position and relative flow through the valve is denoted as flow characteristic. When flow at all valve opening positions is taken at constant inlet pressure ion, the flow characteristic thus determined is referred to as inherent. Figure 2-9 shows such inherent flow characteristics that are typical for flow control valves.

In the most practical applications, however, the pressure loss through the valve varies with valve opening position. This is illustrated in Figure 2-10 for a flow system incorporating a pump. The upper portion of the figure represents the pump characteristic, displaying flow against pump pressure, and the system characteristic, displaying flow against pipeline pressure loss. The lower portion of the figure shows the flow rate against valve opening position. The latter characteristic is referred to as the installed valve flow characteristic and is unique for each valve installation. When the valve has been opened further to increase the flow rate, the pressure at the inlet of the valve decreases, as shown in Figure 2-10. The required rate of valve opening is, therefore, higher in this case than indicated by the inherent flow characteristic.

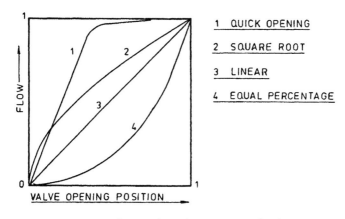

1 QUICK OPENING

2 SQUARE ROOT

3 LINEAR

4 EQUAL PERCENTAGE

Figure 2-9. Inherent Flow Characteristics of Valves.

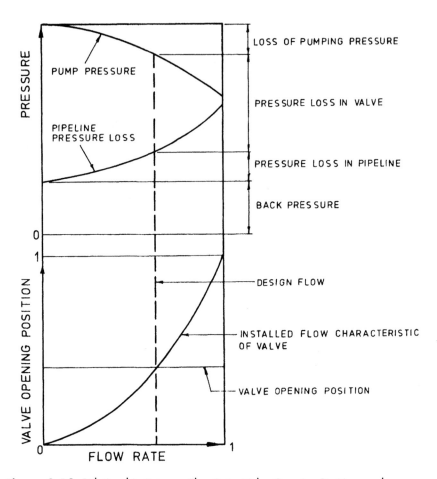

Figure 2-10. Relationship Between Flow Rate, Valve Opening Position, and Pressure Loss in a Pumping system.

If the pump and system characteristic shows that the valve has to absorb a high-pressure drop, the valve should be sized so that the required pressure drop does not occur near the closed position, since this will promote damage to the seatings from the flowing fluid. This consideration leads frequently to a valve size smaller than the adjoining pipe.

Cavitation of Valves

When a liquid passes through a partially closed valve, the static pressure in the region of increasing velocity and in the wake of the closure member drops and may reach the vapor pressure of the liquid. The liquid in the low-pressure region then begins to vaporize and form vapor-filled cavities, which grow around minute gas bubbles and impurities carried by the liquid. When the liquid again reaches a region of high static pressure, the vapor bubbles collapse suddenly or implode. This process is called cavitation.

The impinging of the opposing liquid particles of the collapsing vapor bubble produces locally high but short-lived pressures. If the implosions occur at or near the boundaries of the valve body or the pipe wall, the pressure intensities can match the tensile strength of these parts. The rapid stress reversals on the surface and the pressure shocks in the pores of the boundary surface lead finally to local fatigue failures that cause the boundary surface to roughen until, eventually, quite large cavities form.

The cavitation performance of a valve is typical for a particular valve type, and it is customarily defined by a cavitation index, which indicates the degree of cavitation or the tendency of the valve to cavitate. This parameter is presented in the literature in various forms. The following is a convenient index used by the United States Bureau of Reclamation.[21,22]

$$C = \frac{P_d - P_v}{P_u - P_d} \qquad (2-25)$$

where

C = cavitation index
P_v = vapor pressure relative to atmospheric pressure (negative)
P_d = pressure in pipe 12 pipe diameters downstream of the valve seat
P_u = pressure in pipe 3 pipe diameters upstream of the valve seat

Figure 2-11 displays the incipient cavitation characteristics of butterfly, gate, globe, and ball valves, based on water as the flow medium. The characteristics have been compiled by the Sydney Metropolitan Water Sewerage and Drainage Board, and are based on laboratory observations and published data.[23] Because temperature entrained air, impurities, model tolerances, and the observer's judgment influence the test results, the graphs can serve only as a guide.

The development of cavitation can be minimized by letting the pressure drop occur in stages. The injection of compressed air immediately downstream of the valve minimizes the formation of vapor bubbles by raising the ambient pressure. On the debit side, the entrained air will interfere with the reading of any downstream instrumentation.

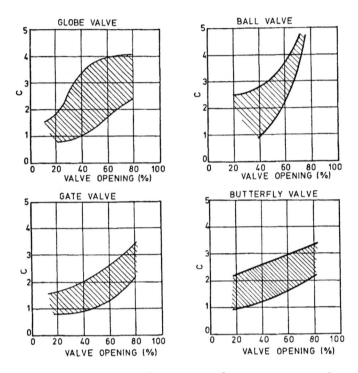

Figure 2-11. Incipient Cavitation Characteristics of Various "In-Line" Valves. (*Courtesy of The Institution of Engineers, Australia.13.*)

A sudden enlargement of the flow passage just downstream of the valve seat can protect the boundaries of valve body and pipe from cavitation damage. A chamber with the diameter of 1.5 times the pipe diameter and a length of 8 times the pipe diameter including the exit taper has proved satisfactory for needle valves used in waterworks.[24]

Waterhammer from Valve Operation

When a valve is being opened or closed to change the flow rate, the change in kinetic energy of the flowing fluid column introduces a transient change in the static pressure in the pipe. In the case of a liquid, this transient change in the static pressure is sometimes accompanied by a shaking of the pipe and a hammering sound—thus the name waterhammer.

The transient pressure change does not occur instantaneously along the entire pipeline but progressively from the point at which the change of flow has been initiated. If, for example, a valve at the end of a pipeline is closed instantaneously, only the liquid elements at the valve feel the valve closure immediately. The kinetic energy stored in the liquid elements then compresses these elements and expands the adjoining pipe walls. The other portion of the liquid column continues to flow at its original velocity until reaching the liquid column which is at rest.

The speed at which the compression zone extends towards the inlet end of the pipeline is uniform and equals the velocity of sound in the liquid within the pipe. When the compression zone has reached the inlet pipe end, all liquid is at rest, but at a pressure above the normal static pressure. The unbalanced pressure now creates a flow in the opposite direction and relieves the rise in the static pressure and the expansion of the pipe wall. When this pressure drop has reached the valve, the whole liquid column is again under normal static pressure, but continues to discharge towards the inlet pipe end so that a wave of subnormal pressure is created, starting at the valve. When this pressure wave has made the round trip, the normal pressure and the original direction of flow are restored. Now the cycle starts again and repeats itself until the kinetic energy of the liquid column is dissipated in friction and other losses.

Joukowsky has shown that instantaneous valve closure raises the static pressure in the pipeline by:

$$\Delta P = av\rho/B \qquad\qquad (2-26)$$

where
ΔP = rise in pressure above normal
v = velocity of a rested flow
a = velocity of pressure wave propagation

$$= \left[\frac{K}{\dfrac{\rho}{B}\left(1 + \dfrac{KDc}{Ee}\right)} \right]^{1/2}$$

ρ = density of the liquid
K = bulk modulus of the liquid
E = Young's modulus of elasticity of the pipe wall material
D = inside diameter of pipe
e = thickness of pipe wall
c = pipe restriction factor ($c = 1.0$ for unrestricted piping)

B (SI units) = 1.0
B (imperial units, fps) = g = 32.174 ft/s^2

In the case of steel piping with a D/e ratio of 35 and water flow, the pressure wave travels at a velocity of approximately 1200 m/s (about 4000 ft/s), and the static pressure increases by 13.5 bar for each 1 m/s, or about 60 lb/in^2 for each 1 ft/s instantaneous velocity change.

If the valve does not close instantaneously but within the time of a pressure-wave round trip of 2L/a, where L is the length of the pipeline, the first returning pressure wave cannot cancel the last outgoing pressure wave, and the pressure rise is the same as if the valve were closed instantaneously. This speed of closure is said to be rapid.

If the valve takes longer to close than 2L/a, the returning pressure waves cancel a portion of the outgoing waves so that the maximum pressure rise is reduced. This speed of closure is said to be slow.

To minimize the formation of unduly high surge pressures from opening or closing valves, stop valves should be operated slowly and in a manner that produces a uniform rate of change of the flow velocity. Check valves, however, are operated by the flowing fluid, and their speed of closure is a function of the valve design and the deceleration characteristic of the retarding fluid column.

If the surge pressure is due to a pump stopping, the calculation of the surge pressure must take into account the pump characteristic and the rate of change of the pump speed after the power supply has been cut off.

If the distance between the check valve and the point of pressure wave reflection is long, and the elevation and the pressure at this point are low, the system tolerates a slow-closing check valve. On the other hand, if the distance between the check valve and the point of reflection is short, and the pressure at this point is high, the flow reverses almost instantaneously and the check valve must be able to close extremely fast. Such nearly instantaneous reverse flow occurs, for example, in multipump installations in which one pump fails suddenly. Guidelines on the selection of check valves for speed of closure are given in Chapter 4.

Calculation of the fluid pressure and velocity as a function of time and location along a pipe can be accomplished in several ways. For simple cases, graphical and algebraic methods can be used. However, the ready availability of digital computers has made the use of numerical methods convenient and allows solutions to any desired accuracy to be obtained. See Reference 25 for a description of this calculating method.

In some cases it may be impossible or impractical to reduce the effects of waterhammer by adjusting the valve characteristic. Consideration should then be given to changing the characteristic of the piping system. One of the most common ways of achieving this is to incorporate one or more surge protection devices at strategic locations in the piping system. Such devices may consist of a standpipe containing gas in direct contact with the liquid or separated from the liquid by a flexible wall or a pressure relief valve.

The effects of waterhammer may also be altered by deliberately changing the acoustic properties of the fluid. This can be done, for example, by introducing bubbles of a non-dissolvable gas directly into the fluid stream. The effect of this is to reduce the effective density and bulk modulus of the fluid. A similar effect can be achieved if the gas is enclosed in a flexible walled conduit, or hose, which runs the length of the pipe.

If even a small amount of gas is present, the effect of pipe wall elasticity in Equation 2-26 becomes insignificant, and the modified acoustic velocity may be expressed by:

$$a = \sqrt{\frac{BK}{\rho}} \qquad\qquad (2-27)$$

where
K = modified fluid bulk modulus

$$= \frac{K_l}{1 + \left(\dfrac{V_g}{V_t}\right)\left(\dfrac{K_l}{K_g} - 1\right)}$$

ρ = modified fluid density

$$= \rho_g \frac{V_g}{V_t} + \rho_l \frac{V_l}{V_t}$$

B = 1.0 (coherent SI units)
B = 32.174 ft/s^2 (imperial units, fps)
K_g = bulk modulus of gas
K_l = bulk modulus of liquid
V_g = volume of gas
V_l = volume of liquid
V_t = total volume
ρ_g = density of gas
ρ_l = density of liquid

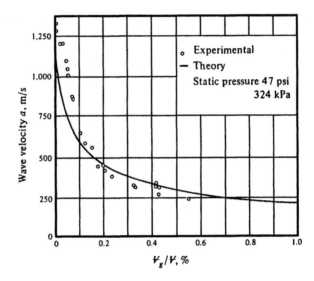

Figure 2-12. Propagation Velocity α of a Pressure Wave in Pipeline for Varying Air Content (Theoretical and Experimental Results).[26] *Reprint from Fluid Transients by E. B. Wylie and V. L Streeter, by courtesy of the authors.*

Figure 2-12 shows the effect of air content on the wave propagation velocity in water. Note that a small amount of air content produces a wave speed less than the speed of sound in air, which, for dry air at 20°C (68°F) and atmospheric pressure, is 318 m/s (1042 ft/s).

Attenuation of Valve Noise[27]

The letting down of gas by valves from a high to a low pressure can produce a troublesome and, in extreme cases, unbearable noise. A major portion of the noise arises from the turbulence generated by the high-velocity jet shearing the relatively still medium downstream of the valve. A silencer found successful in combating this noise is the perforated diffuser, in which the gas is made to flow through numerous small orifices. The diffuser may consist of a perforated flat plate, cone, or bucket.

The diffuser attenuates the low and mid frequencies of the valve noise, but also regenerates a high frequency noise in the perforations, which, however, is more readily attenuated by the passage through the pipe and the air than the lower frequencies. A second beneficial effect of the diffuser is to distribute the flow more evenly, over the cross section of the pipe.

Ingard has shown that the normalized acoustic resistance of a perforated flat plate mounted across the pipe is directly proportional to both the Mach number of the flow through the perforations and the factor

$$\left(\frac{1-\sigma}{\sigma}\right)^2$$

where σ = open area ratio of the perforated plate.[28] Although this cannot be directly related to noise attenuation, it would appear that the Mach number should be as large as possible, and σ as small as possible. For practical purposes, a maximum Mach number of 0.9 is suggested. If the available pressure drop across the diffuser is limited, a Mach number with a lower value may have to be chosen. Practical values for the open area ratio may be taken as between 0.1 and 0.3. Practical values lower than 0.1 may result in an excessively large diffuser, while values higher than 0.3 may result in too low an attenuation.

The peak frequency of the jet noise is also inversely proportional to the diameter of the jet. Therefore, from the point of noise attenuation, the diameter of the perforations should be as small as possible. To avoid the nozzles from becoming blocked, nozzles with a minimum diameter of 5 mm are frequently used.

If the flow velocity in the pipe downstream of the silencer is high, the boundary layer turbulence along the pipe may generate a noise comparable with the attenuated valve noise. Experience suggests that this will not be a problem if the Mach number of the flow in the pipe is kept below about 0.3.

Predicting valve noise and silencer performance is a complex matter. Discussions on these subjects, including the design of silencers, may be found in the References 29, 30, 31, 32 and 33. Further discussions on the generation and radiation of piping noise may be found in References 34 and 35.

3

MANUAL VALVES

FUNCTIONS OF MANUAL VALVES

A manual valve is considered to be a valve that is operated by plant personnel directly, by the use of either a handwheel/wrench or an on/off actuator in the case of shutdown valves. Certain automated valves with actuators are also supplied with handwheels to allow manual operation in the event of power failure.

Manual valves serve three major functions in fluid-handling systems: stopping and starting flow, controlling flow rate, and diverting flow. Valves for stopping and starting flow are frequently employed also for controlling flow rate, and vice versa, while valves for diverting flow are designed for that single purpose.

GROUPING OF VALVES BY METHOD OF FLOW REGULATION

Manual valves may be grouped according to the way the closure member moves onto the seat. Four groups of valves are thereby distinguishable:

1. Closing-down valves. A stopper-like closure member is moved to and from the seat in the direction of the seat axis.

2. Slide valves. A gate-like closure member is moved across the flow passage.
3. Rotary valves. A plug or disc-like closure member is rotated within the flow passage, around an axis normal to the flow stream.
4. Flex-body valves. The closure member flexes the valve body.

Each valve group represents a number of distinct types of valves that use the same method of flow regulation, but differ in the shape of the closure member. For example, plug valves and butterfly valves are both rotary valves, but of a different type. In addition, each valve is made in numerous variations to satisfy particular service needs. Figure 3-1 illustrates the principal methods of flow regulation and names the types of valves, which belong to a particular valve group.

SELECTION OF VALVES

The method by which the closure member regulates the flow and the configuration of the flow path through the valve impart a certain flow characteristic to the valve, which is taken into account when selecting a valve for a given flow-regulating duty.

Valves for Stopping and Starting Flow

These valves are normally selected for low flow resistance and a straight-through flow passage. Such valves are slide valves, rotary valves, and flex-body valves. Closing-down valves offer by their tortuous flow passage a higher flow resistance than other valves and are therefore less frequently used for this purpose. However, if the higher flow resistance can be accepted—as is frequently the case—closing-down valves may likewise be used for this duty.

Valves for Controlling Flow Rate

These are selected for easy adjustment of the flow rate. Closing-down valves lend themselves for this duty because of the directly proportional relationship between the size of the seat opening and the travel of the closure member. Rotary valves and flex-body valves also offer good throttling control, but normally only over a restricted valve-opening range.

Groups of Valves By Method of Flow Regulation	Valve Types	Refer to Page
Closing Down	Globe Valve	54
	Piston Valve	69
Sliding	Parallel Gate Valve	73
	Wedge Gate Valve	84
Rotating	Plug Valve	97
	Ball Valve	108
	Butterfly Valve	120
Flexing of Valve Body	Pinch Valve	137
	Diaphragm Valve	143

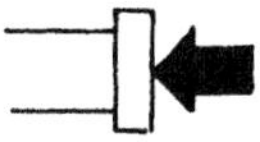

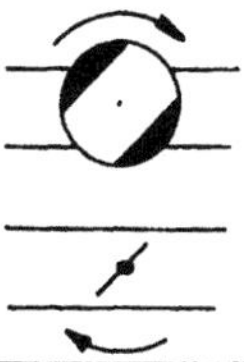

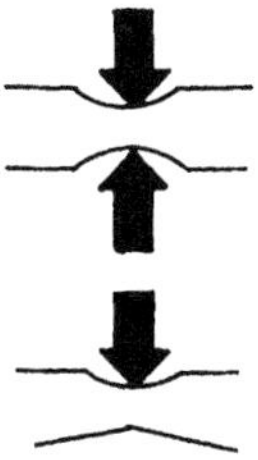

Figure 3-1. Principal Types of Valves Grouped According to the Method of Flow Regulation.

Gate valves, in which a circular disc travels across a circular seat opening, achieve good flow control only near the closed valve position and are therefore not normally used for this duty.

Valves for Diverting Flow

These valves have three or more ports, depending on the flow diversion duty. Valves that adapt readily to this duty are plug valves and ball valves.

For this reason, most valves for the diversion of flow are one of these types. However, other types of valves have also been adapted for the diversion of flow, in some cases by combining two or more valves that are suitably interlinked.

Valves for Fluids with Solids in Suspension

If the fluid carries solids in suspension, the valves best suited for this duty have a closure member that slides across the seat with a wiping motion. Valves in which the closure member moves squarely on and off the seat may trap solids and are therefore suitable only for essentially clean fluids unless the seating material can embed trapped solids.

Valve End Connections

Valves may be provided with any type of end connection used to connect piping. The most important of these for valves are threaded, flanged, and welding end connections.

Threaded end connections. These are made, as a rule, with taper or parallel female threads, which screw over tapered male pipe threads. Because a joint made up in this way contains large leakage passages, a sealant or filler is used to close the leakage passages. If the construction material of the valve body is weldable, screwed joints may also be seal welded. If the mating parts of the joint are made of different but weldable materials with widely differing coefficients of expansion, and if the operating temperature cycles within wide limits, seal welding the screwed joint may be necessary.

Valves with threaded ends are primarily used in sizes up to DN 50 (NPS 2). As the size of the valve increases, installing and sealing the joint become rapidly more difficult. Threaded end valves are available, though, in sizes up to DN 150 (NPS 6).

To facilitate the erection and removal of threaded end valves, couplings are used at appropriate points in the piping system. Couplings up to DN 50 (NPS 1) consist of unions in which a parallel thread nut draws two coupling halves together. Larger couplings are flanged.

Codes may restrict the use of threaded end valves, depending on application.

Flanged end connections. These permit valves to be easily installed and removed from the pipeline. However, flanged valves are bulkier than threaded end valves and correspondingly dearer. Because flanged joints are tightened by a number of bolts, which individually require less tightening torque than a corresponding screwed joint, they can be adapted for all sizes and pressures. At temperatures above 350°C (660°F), however, creep relaxation of the bolts, gasket, and flanges can, in time, noticeably lower the bolt load. Highly stressed flanged joints can develop leakage problems at these temperatures.

Flange standards may offer a variety of flange face designs and also recommend the appropriate flange face finish. As a rule, a serrated flange face finish gives good results for soft gaskets. Metallic gaskets require a finer flange finish for best results. Chapter 2 discusses the design of gaskets.

Welding end connections. These are suitable for all pressures and temperatures, and are considerably more reliable at elevated temperatures and other severe applications than flanged connections. However, removal and re-erection of welding end valves is more difficult. The use of welding end valves is therefore normally restricted to applications in which the valve is expected to operate reliably for long periods, or applications which are critical or which involve high temperatures.

Welding end valves up to DN 50 (NPS 2) are usually provided with welding sockets, which receive plain end pipes. Because socket weld joints form a crevice between socket and pipe, there is the possibility of crevice corrosion with some fluids. Also, pipe vibrations can fatigue the joint. Therefore, codes may restrict the use of welding sockets.

Standards Pertaining to Valve Ends

Appendix C provides a list of the most important U.S. and British standards pertaining to valve ends.

Valve Ratings

The rating of valves defines the pressure-temperature relationship within which the valve may be operated. The responsibility for determining valve ratings has been left over the years largely to the individual manufacturer. The frequent U.S. practice of stating the pressure rating of general purpose valves in terms of WOG (water, oil, gas) and WSP (wet steam pressure) is a carryover from the days when water, oil, gas, and wet steam were the substances generally carried in piping systems. The WOG rating refers to the room-temperature rating, while the WSP rating is usually the high temperature rating. When both a high and a low temperature rating is given, it is generally understood that a straight-line pressure-temperature relationship exists between the two points.

Some U.S. and British standards on flanged valves set ratings that equal the standard flange rating. Both groups of standards specify also the permissible construction material for the pressure-containing valve parts. The rating of welding end valves corresponds frequently to the rating of flanged valves. However, standards may permit welding end valves to be designed to special ratings that meet the actual operating conditions.[36] If the valve contains components made of polymeric materials, the pressure-temperature relationship is limited, as determined by the properties of the polymeric material. Some standards for valves containing such materials as ball valves specify pressure-temperature limits for the valve. Where such standards do not exist, it is the manufacturer's responsibility to state the pressure and temperature limitations of the valve.

Valve Selection Chart

The valve selection chart shown in Table 3-1 is based on the foregoing guidelines and may be used to select the valve for a given flow-regulating duty. The construction material of the valve is determined on one hand by the operating pressure and temperature in conjunction with the applicable valve standard, and on the other hand by the properties of the fluid, such as its corrosive and erosive properties. The selection of the seating material is discussed in Chapter 2. The best approach is to consult the valve manufacturer's catalog, which usually states the standard construction material for the valve for given operating conditions. However, it is the purchaser's responsibility to ensure that the construction material of the valve is compatible with the fluid to be handled by the valve.

If the cost of the valve initially selected is too high for the purpose of the valve, the suitability of other types of valves must be investigated. Sometimes a compromise has to be made.

This selection procedure requires knowledge of the available valve types and their variations, and possibly knowledge of the available valve standards. This information is given in the following sections.

Table 3-1
Valve Selection Chart

Valve Group	Type	Mode of Flow Regulation			Fluid				
		On-Off	Throttling	Diverting	Free of solids	Solids in Suspension non-abrasive	abrasive	Sticky	Sanitary
Closing down	Globe:								
	–straight pattern	Yes	Yes		Yes				
	–angle pattern	Yes	Yes		Yes	special	special		
	–oblique pattern	Yes	Yes		Yes	special	special		
	–multiport pattern			Yes	Yes				
	Piston	Yes	Yes		Yes	Yes	special		
Sliding	Parallel gate:								
	–conventional	Yes			Yes				
	–conduit gate	Yes			Yes	Yes	Yes		
	–knife gate	Yes	special		Yes	Yes	Yes		
	Wedge gate:								
	–with bottom cavity	Yes			Yes				
	–without bottom cavity (rubber seated)	Yes	moderate		Yes	Yes			
Rotating	Plug:								
	–non-lubricated	Yes	moderate	Yes	Yes	Yes			Yes
	–lubricated	Yes	Yes	Yes	Yes	Yes	Yes		
	–eccentric plug	Yes	moderate	Yes	Yes	Yes		Yes	
	–lift plug	Yes		Yes	Yes	Yes		Yes	
	Ball	Yes	moderate	special	Yes	Yes			
	Butterfly	Yes	Yes	special	Yes	Yes			Yes
Flexing	Pinch	Yes	Yes	special	Yes	Yes	Yes	Yes	Yes
	Diaphragm:								
	–weir type	Yes	Yes		Yes	Yes		Yes	Yes
	–straight-through	Yes	moderate		Yes	Yes		Yes	Yes

GLOBE VALVES

Globe valves are closing-down valves in which the closure member is moved squarely on and off the seat. It is customary to refer to the closure member as a disc, irrespective of its shape.

By this mode of disc travel, the seat opening varies in direct proportion to the travel of the disc. This proportional relationship between valve opening and disc travel is ideally suited for duties involving regulation of flow rate. In addition, the seating load of globe valves can be positively controlled by a screwed stem, and the disc moves with little or no friction onto the seat, depending on the design of seat and disc. The sealing capacity of these valves is therefore potentially high. On the debit side, the seatings may trap solids, which travel in the flowing fluid.

Globe valves may, of course, be used also for on-off duty, provided the flow resistance from the tortuous flow passage of these valves can be accepted. Some globe valves are also designed for low flow resistance for use in on-off duty. Also, if the valve has to be opened and closed frequently, globe valves are ideally suited because of the short travel of the disc between the open and closed positions, and the inherent robustness of the seatings to the opening and closing movements.

Globe valves may therefore be used for most duties encountered in fluid-handling systems. This wide range of duties has led to the development of numerous variations of globe valves designed to meet a particular duty at the lowest cost. The valves shown in Figure 3-2 through Figure 3-11 are representative of the many variations that are commonly used in pipelines for the control of flow. Those shown in Figure 3-12 through Figure 3-14 are specialty valves, which are designed to meet a special duty, as described in the captions of these illustrations.

An inspection of these illustrations shows numerous variations in design detail. These are discussed in the following section.

Valve Body Patterns

The basic patterns of globe-valve bodies are the standard pattern, as in the valves shown in Figure 3-2, Figure 3-3, Figure 3-5, Figure 3-6, Figure 3-9 and Figure 3-11; the angle pattern, as in the valves shown in Figure 3-4 and

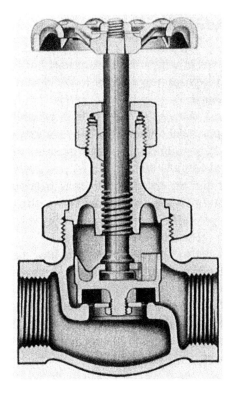

Figure 3-2. Globe Valve, Standard Pattern, Union Bonnet, Internal Screw and Renewable Soft Disc. (*Courtesy of Crane Co.*)

Figure 3-8; and the oblique pattern, as in the valves shown in Figure 3-7 and Figure 3-10, Figure 3-12 and Figure 3-13.

The standard-pattern valve body is the most common one, but offers by its tortuous flow passage the highest resistance to flow of the patterns available.

If the valve is to be mounted near a pipe bend, the angle-pattern valve body offers two advantages. First, the angle-pattern body has a greatly reduced flow resistance compared to the standard-pattern body. Second, the angle-pattern body reduces the number of pipe joints and saves a pipe elbow.

The oblique pattern globe-valve body is designed to reduce the flow resistance of the valve to a minimum. This is particularly well achieved in the valve shown in Figure 3-10. This valve combines low flow resistance for on-off duty with the robustness of globe-valve seatings.

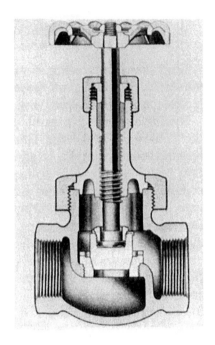

Figure 3-3. Globe Valve, Standard Pattern, Union Bonnet, Internal Screw, Plug Disc. (*Courtesy of Crane Co.*)

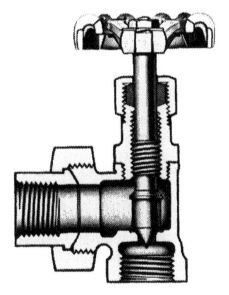

Figure 3-4. Globe Valve, Angle Pattern, Screwed-in Bonnet, Internal Screw Needle Disc. (*Courtesy of Crane Co.*)

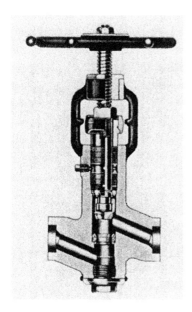

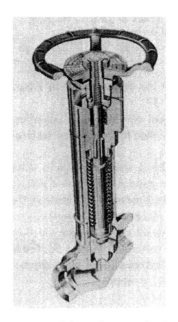

Figure 3-5. Globe Valve, Standard Pattern, Integral Bonnet, External Screw, Plug Disc. (*Courtesy of Velan Engineering Limited*)

Figure 3-6. Globe Valve, Standard Pattern, Welded Bonnet, External Screw, Plug Disc, Bellows Stem Seal with Auxiliary Compression packing. (*Courtesy of Pegler Hattersley Limited*)

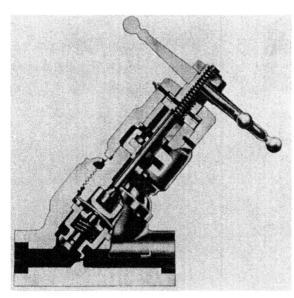

Figure 3-7. Globe Valve, Oblique Pattern, Screwed-in and Seal-Welded Bonnet, External Screw, Plug Disc, with Domed Diaphragm Stem Seal and Auxiliary Compression Packing for Nuclear Application. (*Courtesy of Edward Valves Inc.*)

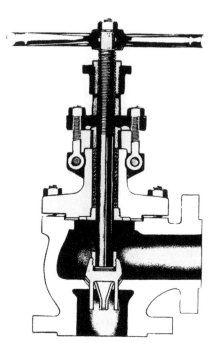

Figure 3-8. Globe Valve, Angle Pattern, Bolted Bonnet, External Screw, Plug Disc with V-Port Skirt for Sensitive Throttling Control. (*Courtesy of Crane Co.*)

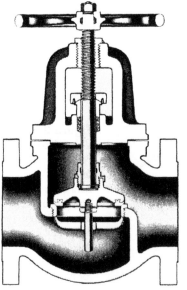

Figure 3-9. Globe Valve, Standard Pattern, Bolted Bonnet, External Screw, Plug Disc. (*Courtesy of Crane Co.*)

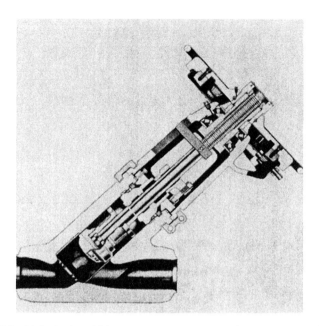

Figure 3-10. Globe Valve, Oblique Pattern, Pressure-Seal Bonnet, External Screw, with Impact Handwheel, Plug Disc. (*Courtesy of Edward Valves Inc.*)

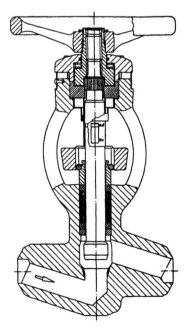

Figure 3-11. Globe Valve, Standard Pattern, Integral Bonnet, Plug-Type Disc Integral with Non-Rotating Stem. (*Courtesy of Sempell A.G.*)

Figure 3-12. Globe Valve, Oblique Pattern, Split Body, External Screw, with Seat-Wiping Mechanism for Application in Slurry Service. (*Courtesy of Langley Alloys Limited.*)

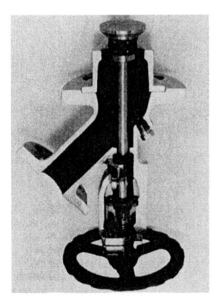

Figure 3-13. Globe Valve Adapted for the Draining of Vessels, Seat Flush with Bottom of Vessel. (*Courtesy of Langley Alloys Limited.*)

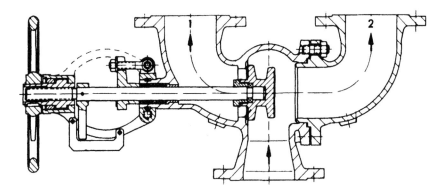

Figure 3-14. Globe Valve, Three-Way, Used as Change-Over Valve in Pressure Relief Valve Installations: One Pressure Relief Valve is Isolated While the Second One Is in Service. (*Courtesy of Bopp & Reuther GmbH.*)

Valve Seatings

Globe valves may be provided with either metal seatings or soft seatings. In the case of metal seatings, the seating stress must not only be high but also circumferentially uniform to achieve the desired degree of fluid tightness. These requirements have led to a number of seating designs. The ones shown in Figure 3-15 are common variations.

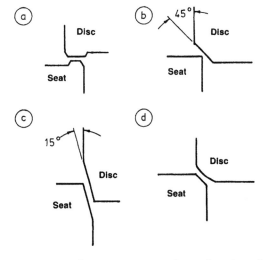

Figure 3-15. Seating Configurations Frequently Employed in Globe Valves.

Flat seatings (see Figure 3-15a) have the advantage over other types of seatings in that they align readily to each other without having to rely on close guiding of the disc. Also, if the disc is moved onto the seat without being rotated, the seatings mate without friction. The resistance of the seating material to galling is therefore unimportant in this case. Deformation of the roundness of the seat due to pipeline stresses does not interfere with the sealability of the seatings as long as the seat face remains flat. If flow is directed from above the seat, the seating faces are protected from the direct impact of solids or liquid droplets travelling in the fluid.

By tapering the seatings, as shown in Figure 3-15b, c, and d, the seating stress for a given seating load can be greatly increased. However, the seating load can be translated into higher uniform seating stress only if the seatings are perfectly mated; that is, they must not be mated with the disc in a cocked position. Thus, tapered discs must be properly guided into the seat. Also, the faces of seat and disc must be perfectly round. Such roundness is sometimes difficult to maintain in larger valves where pipeline stresses may be able to distort the seat roundness. Furthermore, as the seatings are tightened, the disc moves further into the seat. Tapered seatings therefore tighten under friction even if the disc is lowered into the seat without being rotated. Thus the construction material for seat and disc must be resistant to galling in this case.

The tapered seatings shown in Figure 3-15b have a narrow contact face, so the seating stress is particularly high for a given seating load. However, the narrow seat face is not capable of guiding the disc squarely into the seat to achieve maximum sealing performance. But if the disc is properly guided, such seatings can achieve an extremely high degree of fluid tightness. On the debit side, narrow-faced seatings are more readily damaged by solids or liquid droplets than wide-faced seatings, so they are used mainly for gases free of solids and liquid droplets.

To improve the robustness of tapered seatings without sacrificing seating stress, the seatings shown in Figure 3-15c are tapered and provided with wide faces, which more readily guide the disc into the seat. To achieve a high seating stress, the seat face in initial contact with the disc is relatively narrow, about 3 mm ($\frac{1}{8}$ in.) wide. The remainder of the seat-bore is tapered slightly steeper. As the seating load increases, the disc slips deeper into the seat, thereby increasing the seating width. Seatings designed in this way are not as readily damaged by erosion as the seatings in Figure 3-15b. In addition, the long taper of the disc improves the throttling characteristic of the valve.

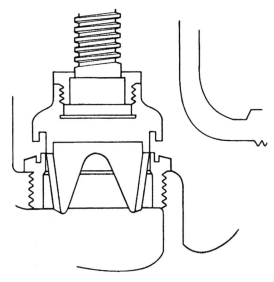

Figure 3-16. Seatings of Globe Valves Adapted for Throttling Duty. (*Courtesy of Pegler Hattersley Limited.*)

The performance of such seatings may be improved by hollowing out the disc to impart some elasticity to the disc shell, as is done in the valve shown in Figure 3-11. This elasticity permits the disc to adapt more readily to deviations of the seatings from roundness.

The seatings shown in Figure 3-15d are ball shaped at the disc and tapered at the seat. The disc can therefore roll, to some extent, on the seat until seat and disc are aligned. Because the contact between the seatings approaches that of a line, the seating stress is very high. On the debit side, the line contact is prone to damage from erosion. The ball-shaped seatings are therefore used only for dry gases, which are also free of solids. This construction is used mainly by U.S. manufacturers.

If the valve is required for fine throttling duty, the disc is frequently provided with a needle-shaped extension, as in the valve shown in Figure 3-4; or with a V-port skirt, as in the valve shown in Figure 3-8 and in the seatings shown in Figure 3-16. In the latter design, the seating faces separate before the V-ports open. The seating faces are, in this way, largely protected against erosion.

An example of soft seating design is the valve shown in Figure 3-2. The soft seating insert is carried in this case by the disc, and may be renewed readily.

Connection of Disc to Stem

The stem of a globe valve may be designed to rotate while raising or lowering the disc, or be prevented from rotating while carrying out this task. These modes of stem operation have a bearing on the design of the disc-to-stem connection.

Most globe valves incorporate a rotating stem because of simplicity of design. If the disc is an integral component of the stem in this case, as it frequently is in small needle valves such as those shown in Figure 3-4, the seatings will mate while the disc rotates, possibly resulting in severe wear of the seatings. Therefore, the main field of application of such valves is for regulating duty with infrequent shut-off duty. For all other duties involving rotating stems, the disc is designed to swivel freely on the stem. However, swivel discs should have minimum free axial play on the stem to prevent the possibility of rapid reciprocating movements of the disc on the stem in the near closed valve position. Also, if the disc is guided by the stem, there should be little lateral play between stem and disc to prevent the disc from landing on the seat in a cocked position.

In the case of nonrotating stems, as in the valves shown in Figures 3-6, Figure 3-7, Figure 3-10, and Figure 3-11, the disc may be either an integral part of the stem (see Figure 3-11) or a separate component from the stem (see Figure 3-6, Figure 3-7, and Figure 3-10). Nonrotating stems are required in valves with diaphragm or bellows valve stem seal, as in Figure 3-6 and Figure 3-7. They are also used in high pressure valves such as those shown in Figure 3-10 and Figure 3-11 to facilitate the incorporation of power operators.

Inside and Outside Stem Screw

The screw for raising or lowering the stem may be located inside the valve body, as in the valves shown in Figure 3-2 through Figure 3-4, or outside the valve body, as in the valves shown in Figure 3-5 through Figure 3-14.

The inside screw permits an economical bonnet construction, but it has the disadvantage that it cannot be serviced from the outside. This construction is therefore best suited for fluids that have good lubricity. For the majority of minor duties, however, the inside screw gives good service.

The outside screw can be serviced from the outside and is therefore preferred for severe duties.

Bonnet Joints

Bonnets may be joined to the valve body by screwing, flanging, welding, or by means of a pressure-seal mechanism; or the bonnet may be an integral part of the valve body.

The screwed-in bonnet found in the valve shown in Figure 3-4 is one of the simplest and least expensive constructions. However, the bonnet gasket must accommodate itself to rotating faces, and frequent unscrewing of the bonnet may damage the joint faces. Also, the torque required to tighten the bonnet joint becomes very large for the larger valves. For this reason, the use of screwed-in bonnets is normally restricted to valve sizes not greater than ND 80 (NPS 3).

If the bonnet is made of a weldable material, the screwed-in bonnet may be seal welded, as in the valves shown in Figure 3-6 and Figure 3-7, or the bonnet connection may be made entirely by welding. These constructions are not only economical but also most reliable irrespective of size, operating pressure, and temperature. On the debit side, access to the valve internals can be gained only by removing the weld. For this reason, welded bonnets are normally used only where the valve can be expected to remain maintenance-free for long periods, where the valve is a throw-away valve, or where the sealing reliability of the bonnet joint outweighs the difficulty of gaining access to the valve internals.

The bonnet may also be held to the valve body by a separate screwed union ring, as in the valves shown in Figure 3-2 and Figure 3-3. This construction has the advantage of preventing any motion between the joint faces as the joint is being tightened. Repeatedly unscrewing the bonnet, therefore, cannot readily harm the joint faces. As with the screwed-in bonnet, the use of bonnets with a screwed union ring is restricted to valve sizes normally not greater than DN 80 (NPS 3).

Flanged bonnet joints such as those found in the valves shown in Figure 3-8 and Figure 3-9 have the advantage over screwed joints in that the tightenening effort can be spread over a number of bolts. Flanged joints may therefore be designed for any valve size and operating pressure. However, as the valve size and operating pressure increase, the flanged joint becomes increasingly heavy and bulky. Also, at temperatures above 350°C (660°F), creep relaxation can, in time, noticeably lower the bolt load. If the application is critical, the flanged joint may be seal welded.

The pressure-seal bonnet found in the valve shown in Figure 3-10 overcomes this weight disadvantage by letting the fluid pressure tighten the joint. The bonnet seal therefore becomes tighter as the fluid pressure increases. This construction principle is frequently preferred for large valves operating at high pressures and temperatures.

Small globe valves may avoid the bonnet joint altogether, as in the valve shown in Figure 3-5 and Figure 3-11. Access to the valve internals is through the gland opening, which is large enough to pass the valve components.

Stuffing Boxes and Back Seating

Figure 3-17 and Figure 3-19 show three types of stuffing boxes, which are typical for valves with a rising stem.

The stuffing box shown in Figure 3-17 is the basic type in which an annular chamber contains the packing between the gland at the top and a shoulder at the bottom. The underside of the stuffing box carries a back seat which, in conjunction with a corresponding seat around the stem, is used to isolate the packing from the system fluid when the valve is fully open.

The stuffing box shown in Figure 3-18 is supplemented with a condensing chamber at the bottom. The condensing chamber served originally as a cooling chamber for condensable gases such as steam. In this particular case, the condensing chamber has a test plug, which may be removed to test the back seat for leak tightness.

A third variation of the stuffing box has a lantern ring mounted between two packing sections, as shown in Figure 3-19. The lantern ring is used

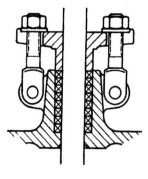

Figure 3-17. Basic Stuffing Box. (*Courtesy of Babcock-Persta Armaturen-Vertriebsgesellschaft mbH.*)

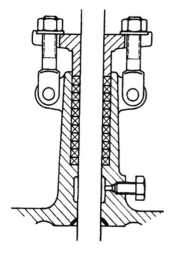

Figure 3-18. Stuffing Box with Condensing Chamber. (*Courtesy of Babcock-Persta Armaturen-Vertriebsgesellschaft mbH.*)

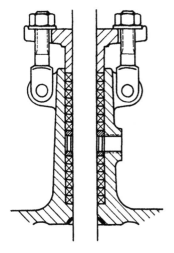

Figure 3-19. Stuffing Box with Lantern Ring. (*Courtesy of Babcock-Persta Armaturen-Vertriebsgesellschaft mbH.*)

mainly in conjunction with compression packings and may serve four different purposes:

1. As an injection chamber for a sealant or an extruded or leached-out lubricant.
2. As a pressure chamber in which an external fluid is pressurized to a pressure equal to or slightly higher than the system pressure to

prevent any leakage of the system fluid to the outside. The external fluid must thereby be compatible with the system fluid and harmless to the surroundings of the valve.

3. As a sealant chamber in vacuum service into which an external fluid is fed to serve as a sealant.
4. As a leakage collection chamber from which the leakage is piped to a safe location.

The inclusion of the lantern ring, however, increases the depth of the packing column. Sidewall friction reduces the gland packing input load as the packing depth increases, leading to an impairment of the seal integrity.[37] Replacing the lantern ring with a spring is used in rotating pump shaft seals to improve the seal integrity.[38]

Direction of Flow Through Globe Valves

The question of direction of flow through globe valves has two answers.

If the possibility exists that flow from above the disc can remove either the disc from the stem or a component from the disc, flow directed from below the disc is mandatory. In this case, hand-operated globe valves with rotating stem and metal seatings can be closed fluid-tight without undue effort, only if the fluid load on the underside of the disc does not exceed about 40–60 kN (9,000–13,000 lb).[39] With a non-rotating stem and roller-bearing supported stem nut, as in the valves shown in Figure 3-10 and Figure 3-11, hand operated globe valves with metal seatings may be closed fluid-tight against a fluid load of about 70–100 kN (16,000–22,000 lb), depending on the leakage criterion and the construction of the valve.[39] One particular advantage of flow directed from below the disc is that the stuffing box of the closed valve is relieved from the upstream pressure. On the debit side, if the valve has been closed against a hot fluid such as steam, thermal contraction of the stem after the valve has been closed can be just enough to induce seat leakage.

If flow is directed from above the disc, the closing force from the fluid acting on top of the disc supplements the closing force from the stem. Thus, this direction of flow increases greatly the sealing reliability of the valve. In this case, hand-operated globe valves with a rotating stem may be opened without excessive effort, only if the fluid load acting on top of the disc dose not exceed about 40–60 kN (9,000–13,000 lb).[39]

If the stem is of the nonrotating type with a roller-bearing supported stem nut, the globe valve may be opened by hand against a fluid load of about 70–100 kN (16,000–22,000 lb).[39] If the fluid load on top of the disc is higher, a bypass valve may have to be provided that permits the downstream system to be pressurized before the globe valve is opened.

Standards Pertaining to Globe Valves

Appendix C provides a list of U.S. and British standards pertaining to globe valves.

Applications

Duty:
    Controlling flow
    Stopping and starting flow
    Frequent valve operation
Service:
    Gases essentially free of solids
    Liquids essentially free of solids
    Vacuum
    Cryogenic

PISTON VALVES

Piston valves are closing-down valves in which a piston-shaped closure member intrudes into or withdraws from the seat bore, as in the valves shown in Figure 3-20 through Figure 3-24.

In these valves, the seat seal is achieved between the lateral faces of the piston and the seat bore. When the valve is being opened, flow cannot start until the piston has been completely withdrawn from the seat bore. Any erosive damage occurs, therefore, away from the seating surfaces. When the valve is being closed, the piston tends to wipe away any solids, which might have deposited themselves on the seat. Piston valves may thus handle fluids that carry solids in suspension. When some damage occurs to the seatings, the piston and the seat can be replaced *in situ*, and the valve is like new without any machining.

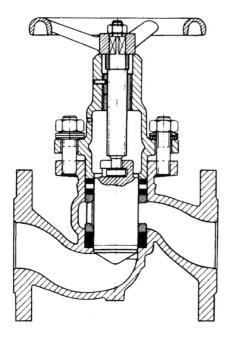

Figure 3-20. Piston Valve, Standard Pattern, Seat Packing Mounted in Valve Body, Piston Pressure Unbalanced. (*Courtesy of Rich. Klinger AG.*)

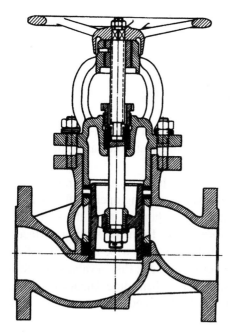

Figure 3-21. Piston Valve, Standard Pattern, Seat Packing Mounted in Valve Body, Piston Pressure Balanced. (*Courtesy of Rich. Klinger AG.*)

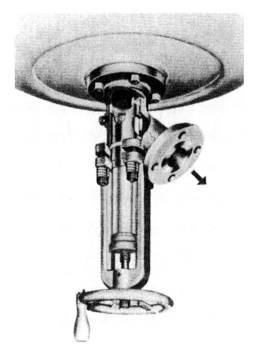

Figure 3-22. Piston Valve, Adapted for Draining Vessels, Seat Packing Mounted in Valve Body. (*Courtesy of Yarway Corporation.*)

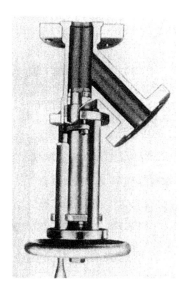

Figure 3-23. Piston Valve Adapted for Draining Vessels, Seat Packing Mounted on Piston; the "Ram-Seal" Principle. (*Courtesy of Fetterolf Corporation.*)

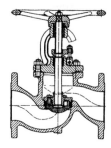

Figure 3-24. Piston Valve, Standard Pattern, Seat Packing Mounted on Piston. (*Courtesy of Rich. Klinger AG.*)

Like globe valves, piston valves permit good flow control. If sensitive flow adjustment is required, the piston may be fitted with a needle shaped extension. Piston valves are also used for stopping and starting flow when flow resistance due to the tortuous flow passage is accepted.

Construction

The seatings of piston valves are formed by the lateral faces of the valve bore and the piston. A fluid-tight contact between these faces is achieved by a packing that either forms part of the valve bore, as in the valves shown in Figure 3-20 through Figure 3-22, or part of the piston, as in the valves shown in Figure 3-23 and Figure 3-24. Packings commonly used for this purpose are compression packings based on compressed asbestos or PTFE and O-ring packings.

In the case of the piston valve shown in Figure 3-20, the piston moves in two packings that are separated by a lantern ring. The lower packing represents the seat packing, while the upper packing seals the piston to the atmosphere. The bonnet serves thereby as the gland that permits both packings to be tightened through tightening of the cover bolts. Disc springs under the nuts of the cover bolts minimize variations in packing stress due to thermal contraction and expansion of the valve parts. When one of the packings leaks, the fluid seal can be restored by retightening the bolts. Retightening must be carried out while the valve is closed to prevent an unrestrained expansion of the seat packing into the valve bore.

The valve shown in Figure 3-21 differs from the one in Figure 3-20 only in that the piston is pressure balanced. The two packings around the piston are both seat packings, and a separate packing is provided for the stem. The purpose of balancing the piston is to minimize the operating effort in large valves operating against high fluid pressures.

The packing train of the valve shown in Figure 3-22 is likewise stressed through the bonnet in conjunction with springs under the nuts of the cover bolts, or with a spring between the bonnet and the packing. However, as the piston moves into the final closing position, a shoulder on the piston contacts a compression ring on top of the packing so that any further progression of the piston tightens the packing still further.

The piston valve shown in Figure 3-23 carries the seat packing on the end of the piston instead of in the valve bore. The packing is supported thereby on its underside by a loose compression ring. When the piston moves into the final closing position, the compression ring comes to rest on a shoulder in the seat bore so that any further progression of the piston causes the compression ring to tighten the packing. Because the packing establishes interference with the seat in the last closing stages only, the operating effort of the valve is lower over a portion of the piston travel than that of the foregoing valves.

The piston valve shown in Figure 3-24 also carries the seat packing on the piston. However, the loose compression ring is replaced by a friction ring that acts as a spring element and, as such, pre-stresses the packing. When the piston moves into the seat, the friction ring comes to rest in the seat bore, and any progression of the piston increases the packing stress.

National standards that apply specifically to piston valves do not exist.

Applications

Duty:
  Controlling flow
  Stopping and starting flow
Service:
  Gases
  Liquids
  Fluids with solids in suspension
  Vacuum

PARALLEL GATE VALVES

Parallel gate valves are slide valves with a parallel-faced gate-like closure member. This closure member may consist of a single disc or twin discs

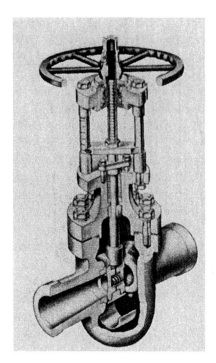

Figure 3-25. Parallel Slide Gate Valve with Converging-Diverging Flow Passage and Follower Eyepiece. (*Courtesy of Hopkinsons Limited.*)

with a spreading mechanism in between. Typical valves of this type are shown in Figure 3-25 through Figure 3-32.

The force that presses the disc against the seat is controlled by the fluid pressure acting on either a floating disc or a floating seat. In the case of twin disc parallel gate valves, this force may be supplemented with a mechanical force from a spreading mechanism between the discs.

One advantage of parallel gate valves is their low resistance to flow, which in the case of full-bore valves approaches that of a short length of straight pipe. Because the disc slides across the seat face, parallel gate valves are also capable of handling fluids, which carry solids in suspension. This mode of valve operation also imposes some limitations on the use of parallel gate valves:

• If fluid pressure is low, the seating force may be insufficient to produce a satisfactory seal between metal-to-metal seatings.

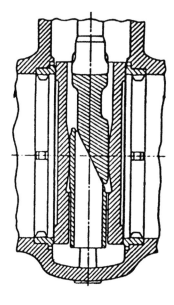

Figure 3-26. Scrap View of Parallel Gate Valve Showing Double-Disc Closure Member with Wedging Mechanism. (*Courtesy of Pacific Valves, Inc.*)

- Frequent valve operation may lead to excessive wear of the seating faces, depending on magnitude of fluid pressure, width of seating faces, lubricity of the fluid to be sealed, and the wear resistance of the seating material. For this reason, parallel gate valves are normally used for infrequent valve operation only.
- Loosely guided discs and loose disc components will tend to rattle violently when shearing high density and high velocity flow.
- Flow control from a circular disc travelling across a circular flow passage becomes satisfactory only between the 50% closed and the fully closed positions. For this reason, parallel gate valves are normally used for on-off duty only, though some types of parallel gate valves have also been adapted for flow control, for example, by V-porting the seat.

The parallel gate valves shown in Figure 3-25 through Figure 3-28 are referred to as conventional parallel gate valves, and those of Figure 3-29 through Figure 3-32 are referred to as conduit gate valves. The latter are full-bore valves, which differ from the former in that the disc seals the valve body cavity against the ingress of solids in both the open and closed valve positions. Such valves may therefore be used in pipelines that have to be scraped.

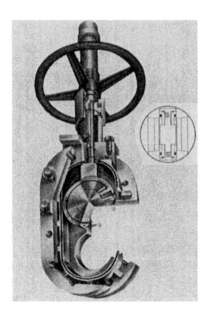

Figure 3-27. Parallel Gate Valve with Scrap View of Seating Arrangement Showing Spring-Loaded Floating Inserts in Disc. (*Courtesy of Grove Valve and Regulator Company.*)

Conventional Parallel Gate Valves

The valves shown in Figure 3-25 through Figure 3-28 are representative of the common varieties of conventional parallel gate valves.

One of the best known is the valve shown in Figure 3-25, commonly referred to as a parallel slide gate valve. The closure member consists of two discs with springs in between. The duties of these springs are to keep the upstream and downstream seatings in sliding contact and to improve the seating load at low fluid pressures. The discs are carried in a belt eye in a manner that prevents their unrestrained spreading as they move into the fully open valve position.

The flow passage of this particular parallel slide gate valve is venturi shaped. The gap between the seats of the fully open valve is bridged by an eyelet to ensure a smooth flow through the valve. The advantages offered by this construction include not only economy of construction but also a reduced operating effort and lower maintenance cost. The only disadvantage is a slight increase in pressure loss across the valve.

The seating stress reaches its maximum value when the valve is nearly closed, at which position the pressure drop across the valve is near

Figure 3-28. Knife Gate Valve. (*Courtesy of DeZurik.*)

maximum; but the seating area in mutual contact is only a portion of the total seating area. As the disc travels between the three-quarter closed to the nearly closed valve position, the flowing fluid tends to tilt the disc into the seat bore, so heavy wear may occur in the seat bore and on the outer edge of the disc. To keep the seating stress and corresponding seating wear within acceptable limits, the width of the seatings must be made appropriately wide. Although this requirement is paradoxical in that the seating width must be small enough to achieve a high seating stress but wide enough to keep seating wear within acceptable limits, the fluid tightness that is achieved by these valves satisfies the leakage criterion of the steam class, provided the fluid pressure is not too low.

Parallel slide gate valves have other excellent advantages: the seatings are virtually self-aligning and the seat seal is not impaired by thermal movements of the valve body. Also, when the valve has been closed in the cold condition, thermal extension of the stem cannot overload the seatings. Furthermore, when the valve is being closed, a high accuracy in the positioning of the discs is not necessary, thus an electric drive for the valve can be travel limited. Because an electric drive of this type is both economical and reliable, parallel slide gate valves are often preferred as block valves in larger power stations for this reason alone. Of course, parallel slide gate valves may be used also for many other services such as water, in particular boiler feed water—and oil.

A variation of the parallel slide gate valve used mainly in the U.S. is fitted with a closure member such as the one shown in Figure 3-26. The closure member consists of two discs with a wedging mechanism in between, which, on contact with the bottom of the valve body, spreads the discs apart. When the valve is being opened again, the wedging mechanism releases the discs. Because the angle of the wedge must be wide enough for the wedge to be self-releasing, the supplementary seating load from the wedging action is limited.

To prevent the discs from spreading prematurely, the valve must be mounted with the stem upright. If the valve must be mounted with the stem vertically down, the wedge must be appropriately supported by a spring.

The performance characteristic attributed to parallel slide gate valves also applies largely to this valve. However, solids carried by the flowing fluid and sticky substances may interfere with the functioning of the wedging mechanism. Also, thermal extension of the stem can overload the seatings. The valve is used mainly in gas, water, and oil services.

Conventional parallel gate valves may also be fitted with soft seatings, as in the valve shown in Figure 3-27. The closure member consists here of a disc that carries two spring-loaded floating seating rings. These rings are provided with a bonded O-ring on the face and a second O-ring on the periphery. When the disc moves into the closed position, the O-ring on the face of the floating seating ring contacts the body seat and produces the initial fluid seal. The fluid pressure acting on the back of the seating ring then forces the seatings into still closer contact.

Because the unbalanced area on the back of the floating rings is smaller than the area of the seat bore, the seating load for a given fluid pressure and valve size is smaller than in the previously described valves. However, the valve achieves a high degree of fluid tightness by means of the O-ring even at low fluid pressures. This sealing principle also permits double block and bleed.

The parallel gate valve shown in Figure 3-28 is known as the knife gate valve, and is designed to handle slurries, including fibrous material. The valve owes its ability to handle these fluids to the knife-edged disc, which is capable of cutting through fibrous material, and the virtual absence of a valve body cavity. The disc travels in lateral guides and is forced against the seat by lugs at the bottom. If a high degree of fluid tightness is required, the valve may also be provided with an O-ring seat seal.

Conduit Gate Valves

Figure 3-29 through Figure 3-32 show four types of valves that are representative of conduit gate valves. All four types of valves are provided with floating seats that are forced against the disc by the fluid pressure.

The seats of the conduit gate valve shown in Figure 3-29 are faced with PTFE and sealed peripherally by O-rings. The disc is extended at the bottom to receive a porthole. When the valve is fully open, the porthole in the disc engages the valve ports so that the disc seals the valve body cavity against the ingress of solids. The sealing action of the floating seats also permits double block and bleed. If the seat seal should fail in service, a temporary seat seal can be produced by injecting a sealant into the seat face.

The conduit gate valve shown in Figure 3-30 differs from the previous one in that the disc consists of two halves with a wedge-shaped

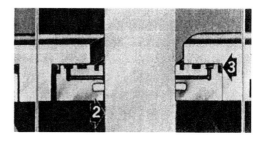

Figure 3-29. Conduit Gate Valve with Scrap View of Seating Arrangement Showing Floating Seats. (*Courtesy of W.K.M. Valve Division, ACF Industries, Inc.*)

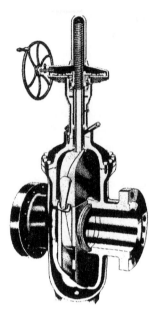

Figure 3-30. Conduit Gate Valve with Floating Seats and Expandable Disc. (*Courtesy of W.K.M. Valve Division, ACF Industries, Inc.*)

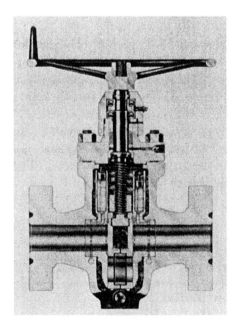

Figure 3-31. Conduit Gate Valve with Automatic Injection of a Sealant to the Downstream Seatings Each Time the Valve Closes. (*Courtesy of McEvoy Oilfield Equipment Company.*)

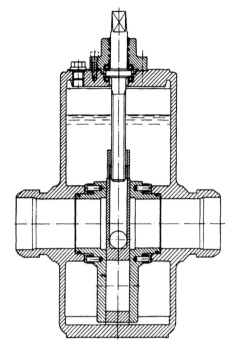

Figure 3-32. Conduit Gate Valve with Metal Seatings and Oil-Filled Body Cavity for Heavily Dust-Laden Gasses and the Hydraulic Transport of Coal and Ores. (*Courtesy of VAG-Armaturen GmbH.*)

interface. These halves are interlinked so that they wedge apart when being moved into the fully open or closed positions, but relax in the intermediate position to permit the disc to travel. Depending on the use of the valve, the face of the floating seats may be metallic or provided with a PTFE insert. To prevent the ingress of solids into the valve body cavity during all stages of disc travel, the floating seats are provided with skirts, between which the disc travels. This valve likewise permits double block and bleed. Also, should the seat seal fail in service, a temporary seat seal can be provided by injecting a sealant into the seat face.

The sealing action of the conduit gate valve shown in Figure 3-31 depends on a sealant that is fed to the downstream seat face each time the valve is operated. For this purpose, the floating seats carry reservoirs that are filled with a sealant and topped by a floating piston. The entire valve chamber is, furthermore, filled with a grease that transmits the fluid pressure to the top of the piston of the downstream reservoir. The closure member consists of two discs, which are spread apart by springs. Both the sealant and the body grease can be replenished from the outside while the valve is in service. Each reservoir filling is sufficient for more than 100 valve operations. This mode of sealing achieves a high degree of fluid tightness at high fluid pressures.

The conduit gate valve shown in Figure 3-32 is especially designed for the hydraulic transport of coal and ore, and the transport of heavily dust-laden gases. The gate, which consists of a heavy wear-resistant plate with a porthole in the bottom, slides between two floating seats that are highly pre-stressed against the disc by means of disc springs. To prevent any possible entry of solids into the body cavity, the seats are provided with skirts for the full travel of the disc. The faces of the disc and seats are metallic and highly polished. The valve chamber is, furthermore, filled with a lubricant that ensures lubrication of the seating faces.

Valve Bypass

The seating load of the larger parallel gate valves (except those with float-ing seats) can become so high at high fluid pressures that friction between the seatings can make it difficult to raise the disc from the closed position. Such valves are therefore frequently provided with a valved bypass line, which is used to relieve the seating load prior to opening the valve. There are no fast rules about when to employ a bypass, and the manufacturer's recommendation may be sought. Some standards of gate valves contain recommendations on the minimum size of the bypass.

In the case of gases and vapors, such as steam, that condense in the cold downstream system, the pressurization of the downstream system can be considerably retarded. In this instance, the size of the bypass line should be larger than the minimum recommended size.

Pressure-Equalizing Connection

In the case of the conventional double-seated parallel gate valves shown in Figure 3-25 and Figure 3-26, thermal expansion of a liquid trapped in the closed valve chamber will force the upstream and downstream discs into more intimate contact with their seats, and cause the pressure in the valve chamber to rise. The higher seating stress makes it in turn more difficult to raise the discs, and the pressure in the valve chamber may quickly become high enough to cause a bonnet flange joint to leak or the valve body to deform. Thus, if such valves are used to handle a liquid with high thermal expansion, they must have a pressure-equalizing connection that connects the valve chamber with the upstream piping.

The pressure rise in the valve chamber may also be caused by the revaporation of trapped condensate, as in the case in which these valves are closed against steam. Both the valve chamber and the upstream piping are initially under pressure and filled with steam. Eventually, the steam will cool, condense, and be replaced to some extent with air.

Upon restart, the steam will enter the upstream piping and, since the upstream seat is not normally fluid-tight against the upstream pressure, will enter the valve chamber. Some of the new steam will also condense initially until the valve body and the upstream piping have reached the saturation temperature of the steam.

When this has happened, the steam begins to boil off the condensate. If no pressure-equalizing connection is provided, the expanding steam will force the upstream and downstream discs into more intimate contact with their seats, and raise the pressure in the valve chamber. The magnitude of the developing pressure is a function of the water temperature and the degree of filling of the valve chamber with water, and may be obtained from Figure 3-33.

The pressure-equalizing connection may be provided by a hole in the upstream disc or by other internal or external means. Some makers of parallel gate valves of the types shown in Figure 3-25 and Figure 3-26 combine the bypass line with a pressure-equalizing line if the valve is intended for steam.

Standards Pertaining to Parallel Gate Valves

Appendix C provides a list of U.S. and British standards pertaining to parallel gate valves.

Applications

Duty:
    Stopping and starting flow
    Infrequent operation
Service:
    Gases

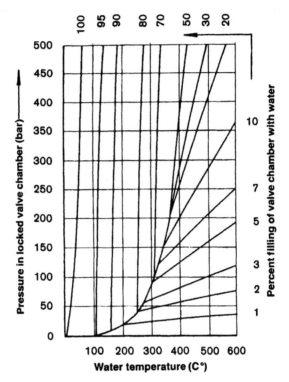

Figure 3-33. Pressure in Locked Valve Chamber as a Result of the Revaporation of Trapped Water Condensate. (*Courtesy of Sempell A.G.*)

Liquids
Fluids with solids in suspension
Knife gate valve for slurries, fibers, powders, and granules
Vacuum
Cryogenic

WEDGE GATE VALVES

Wedge gate valves differ from parallel gate valves in that the closure member is wedge-shaped instead of parallel, as shown in Figure 3-34 through Figure 3-45. The purpose of the wedge shape is to introduce a high supplementary seating load that enables metal-seated wedge gate valves to seal not only against high, but also low, fluid pressures. The degree of seat

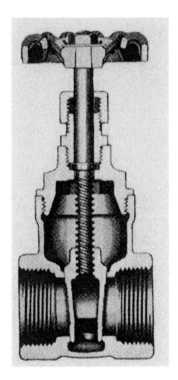 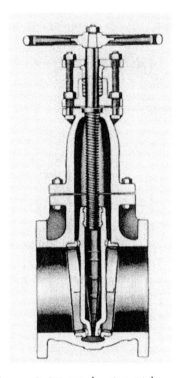

Figure 3-34. Wedge Gate Valve with Plain Hollow Wedge, Screwed-in Bonnet, and Internal Screw. (*Courtesy of Crane Co.*)

Figure 3-35. Wedge Gate Valve with Plain Hollow Wedge, Bolted Bonnet, and Internal Screw. (*Courtesy of Crane Co.*)

tightness that can be achieved with metal-seated wedge gate valves is therefore potentially higher than with conventional metal-seated parallel gate valves. However, the upstream seating load is not normally high enough to permit block-and-bleed operation of metal-to-metal seated wedge gate valves.

The bodies of these valves carry guide ribs, or slots, in which the disc travels. The main purpose of these guides is to carry the wedge away from the downstream seat except for some distance near the closed valve position so as to minimize wear between the seatings. A second purpose of the guides is to prevent the disc from rotating excessively while travelling between the open and closed valve positions. If some rotation occurs, the disc will initially jam on one side between the body seatings and the rotate into the correct position before travelling into the final seating position.

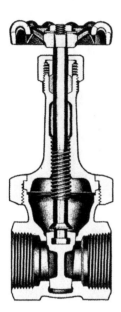

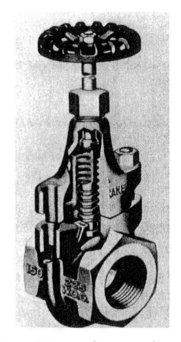

Figure 3-36. Wedge Gate Valve with Plain Solid Wedge, Union Bonnet, and Internal Screw. (*Courtesy of Crane Co.*)

Figure 3-37. Wedge Gate Valve with Clamped Bonnet, Internal Screw. (*Courtesy of Crane Co.*)

There are also types of wedge gate valves that can dispense with a wedge guide, such as the valve shown in Figure 3-45 in which the wedge is carried by the diaphragm.

Compared with parallel gate valves, wedge gate valves also have some negative features:

• Wedge gate valves cannot accommodate a follower conduit as conveniently as parallel gate valves can.
• As the disc approaches the valve seat, there is some possibility of the seatings trapping solids carried by the fluid. However, rubber-seated wedge gate valves, as shown in Figure 3-44 and Figure 3-45, are capable of sealing around small trapped solids.
• An electrical drive for wedge gate valves is more complicated than for parallel gate valves in that the drive must be torque-limited instead of travel-limited. The operating torque of the drive must thereby be high enough to effect the wedging of the wedge into the seats while the valve

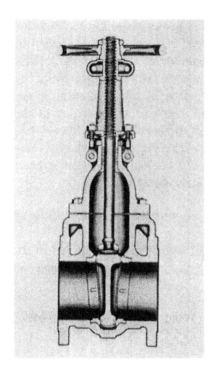

Figure 3-38. Wedge Gate with Plain Solid Wedge, Bolted Bonnet, and External Screw. (*Courtesy of Crane Co.*)

is being closed against the full differential line pressure. If the valve is closed against zero differential pressure, the wedging of the wedge into the seats becomes accordingly higher. To permit the valve to be opened again against the full differential pressure, and to allow also for a possible increase of the operating effort due to thermal movements of the valve parts, the operator must be generously sized.

The limitations of wedge gate valves are otherwise similar to those of parallel gate valves.

Efforts to improve the performance of wedge gate valves led to the development of a variety of wedge designs; the most common ones are described in the following section.

Variations of Wedge Design

The basic type of wedge is the plain wedge of solid or hollow construction, as in the valves shown in Figure 3-34 through Figure 3-39. This design

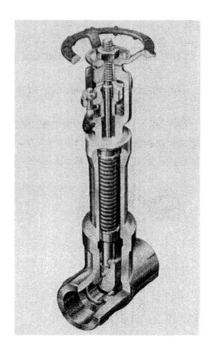

Figure 3-39. Wedge Gate with Plain Solid Wedge, Welded Bonnet, External Screw, Bellows Stem Seal, and Auxiliary Stuffing Box Seal. (*Courtesy of Pegler Hattersley Limited.*)

has the advantage of being simple and robust, but distortions of the valve body due to thermal and pipeline stresses may unseat or jam the metal-seated wedge. A failure of this kind is more often experienced in valves of light-weight construction.

The sealing reliability of gate valves with a plain wedge can be improved by elastomeric or plastic sealing elements in either the seat or the wedge. Figure 3-40 shows a seat in which the sealing element is a PTFE insert. The PTFE insert stands proud of the metal face just enough to ensure a seal against the wedge.

Efforts to overcome the alignment problem of plain wedges led to the development of self-aligning wedges. Figure 3-41, Figure 3-42, and Figure 3-43 show typical examples. The simplest of these is the flexible wedge shown in Figure 3-41 and Figure 3-42, which is composed of two discs with an integral boss in between. The wedge is sufficiently flexible to find its own orientation. Because the wedge is simple and contains no separate components that could rattle loose in service, this construction has become a favored design.

Figure 3-40. Seat of Wedge Gate Valve with PTFE Sealing Insert. (*Courtesy of Crane Co.*)

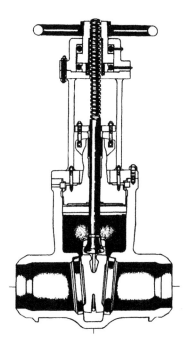

Figure 3-41. Wedge Gate Valve with Flexible Wedge, Pressure-Sealed Bonnet, External Screw. (*Courtesy of Crane Co.*)

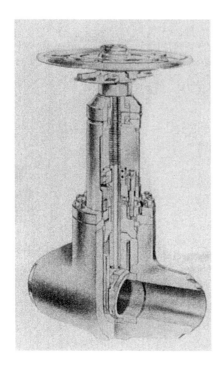

Figure 3-42. Forged Wedge Gate Valve with Pressure-Sealed Bonnet, Incorporating Flexible Wedge with Hard Faced Grooves Sliding on Machined Body Ribs. (*Courtesy of Velan Engineering Ltd.*)

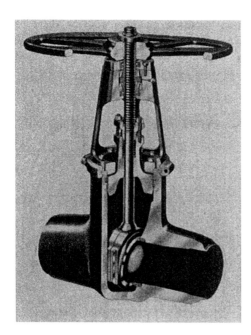

Figure 3-43. Wedge Gate Valve with Two-Piece Wedge, Pressure-Sealed Bonnet, and External Screw. (*Courtesy of Edward Valves Inc.*)

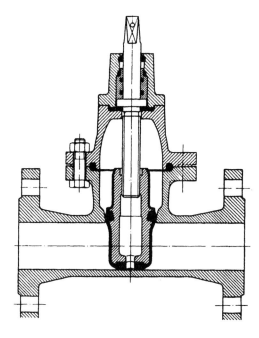

Figure 3-44. Rubber-Seated Wedge Gate Valve without Cavity in Bottom of Valve Body. (*Courtesy of VAG-Armaturen GmbH.*)

The self-aligning wedge of the valve shown in Figure 3-43 consists of two identical tapered plates that rock around a separate spacer ring. This spacer ring may also be used to adjust the wedge assembly for wear. To keep the plates together, the body has grooves in which the wedge assembly travels.

Rubber lining of the wedge, as in the valves shown in Figure 3-44 and Figure 3-45, led to the development of new seating concepts in which the seat seal is achieved in part between the rim of the wedge and the valve body. In this way, it became possible to avoid altogether the creation of a pocket at the bottom of the valve body. These valves are therefore capable of handling fluids carrying solids in suspension, which would otherwise collect in an open body cavity.

In the case of the valve shown in Figure 3-44, the wedge is provided with two stirrup-shaped rubber rings that face the rim of the wedge at the bottom sides, and the top lateral faces. When the valve is being closed, the rubber rings seal against the bottom and the sidewalls of the valve body and, by a wedging action, against the seat faces at the top.

The wedge of the valve shown in Figure 3-45 is completely rubber lined and forms part of a diaphragm, which separates the operating mechanism

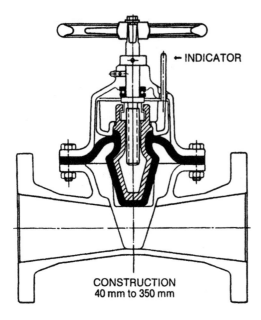

← INDICATOR

CONSTRUCTION
40 mm to 350 mm

Figure 3-45. Rubber-Seated Wedge Gate Valve Without Cavity in Bottom of Valve Body. (*Courtesy of Schmitz & Schulte, D-5093 Burscheid.*)

from the flowing fluid. When the valve is being closed, the bottom of the wedge seals against the bottom of the valve body and the body seats. The valve may also be lined with corrosion-resistant materials, and is therefore widely used in the chemical industry.

Connection of Wedge to Stem

The wedge-to-stem connection usually consists of a T-slot in the top of the wedge, which receives a collar on the stem. According to API standard 600, this connection must be stronger than the weakest stem section so that the wedge cannot become detached from the stem while operating the valve.

The T-slot in the wedge may thereby be oriented in line with the flow passage, as in the valves shown in Figure 3-38, Figure 3-39, and Figure 3-41, or across the flow passage as shown in Figure 3-36 and Figure 3-42. The latter construction permits a more compact valve body design and, therefore, has become popular for economic reasons. Also, this construction favorably lowers the point at which the stem acts on the wedge. However, the T-slot must be wide enough to accommodate the play of the wedge in its guide, allowing also for wear of the guide.

There are also exceptions to this mode of wedge-to-stem connection, as in the valves shown in Figure 3-34 and Figure 3-37, in which the stem must carry the entire thrust on the wedge. For this reason, this construction is suitable for low pressure applications only. In the valve shown in Figure 3-43, guide play is virtually absent, allowing the stem to be captured in the wedge.

Wedge Guide Design

The body guides commonly consist either of ribs, which fit into slots of the wedge, or of slots, which receive ribs of the wedge. Figure 3-42 and Figure 3-43 illustrate these guiding mechanisms.

The body ribs are not normally machined for reason of low cost construction. However, the rough surface finish of such guides is not suited for carrying the travelling wedge under high load. For this reason, the wedge is carried on valve opening initially on the seat until the fluid load has become small enough for the body ribs to carry the wedge. This method of guiding the wedge may require considerable play in the guides, which must be matched, by the play in the T-slot for suspending the wedge on the stem.

Once the body ribs begin to carry the wedge upon valve opening, the wedge must be fully supported by the ribs. If the length of support is insufficient, the force of the flowing fluid acting on the unsupported section of the wedge may be able to tilt the wedge into the downstream seat bore. This support requirement is sometimes not complied with. On the other hand, some valve makers go to any length to ensure full length wedge support.

There is no assurance that the wedge will slide on the stem collar when opening the valve. At this stage of valve operation, there is considerable friction between the contact faces of the T-slot and stem collar, possibly causing the wedge to tilt on the stem as the valve opens. If, in addition, the fit between T-slot and stem collar is tight, and the fluid load on the disc is high, the claws forming the T-slot may crack.

For critical applications, guides in wedge gate valves are machined to close tolerances and designed to carry the wedge over nearly the entire valve travel, as in the valves shown in Figure 3-42 and Figure 3-43.

In the valve shown in Figure 3-42, the wedge grooves are hard-faced and precision-guided on machined guide ribs welded to the valve body. The wedge is permitted in this particular design to be carried by the seat for 5% of the total travel.

In the valve shown in Figure 3-43, the wedge consists of two separate wedge-shaped plates. These carry hard-faced tongues that are guided in machined grooves of the valve body. When wear has taken place in the guides, the original guide tolerance can be restored by adjusting the thickness of a spacer ring between the two wedge plates.

Valve Bypass

Wedge gate valves may have to be provided with bypass connections for the same reason described for parallel gate valves on page 77.

Pressure-Equalizing Connection

In the case of wedge gate valves with a self-aligning double-seated wedge, thermal expansion of a fluid locked in the valve body will force the upstream and downstream seatings into still closer contact and cause the pressure in the valve body cavity to rise. A similar situation may arise with soft-seated wedge gate valves in which the wedge is capable of producing an upstream seat seal. Thus, if such valves handle a liquid with high thermal expansion, or if revaporation of trapped condensate can occur, they may have to be provided with a pressure-equalizing connection as described for parallel gate valves on page 77.

Case Study of Wedge Gate Valve Failure

Figure 3-46 through Figure 3-48 show components of a DN 300 (NPS 12) class 150 wedge gate valve to API standard 600 that failed on first application.

The valve was mounted in a horizontal line under an angle of 45° from the vertical for ease when hand-operating a gear drive. This operating position required the wedge to ride on the body rib. Unfortunately, the guide slot in the wedge had sharp edges. As the wedge traveled on the rib, the sharp edges of the guide slot caught on the rough rib surface, causing the wedge to rotate until contacting the opposite body rib. At this stage, the wedge was jammed. Further closing effort by the valve operator produced the damage to the valve internals shown in Figure 3-46 through Figure 3-48.

Figure 3-46 shows the damage to the wedge guide that was riding on the body rib and Figure 3-47 shows the damaged body rib. Figure 3-48

Figure 3-46. Damage to Guide Slots of Flexible Wedge Resulting from Attempted Closure of Wedge Gate Valve with Wedge Rotated and Jammed in Valve Body, Size DN 300 (NPS 12) Class 150. (*Courtesy of David Mair.*)

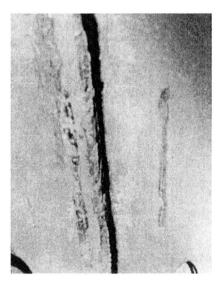

Figure 3-47. Damage to Valve Body Guide Ribs Resulting from Attempted Closure of Wedge Gate Valve with Rotated Wedge Jammed in Valve Body, Size DN 300 (NPS 12) Class 150. (*Courtesy of David Mair.*)

shows the bent valve stem and damage to the stem surface around the stem guide bush.

Further inspection of the valve showed also that play in the wedge guides was larger than the possible travel of the wedge on the stem. Thus, the stem had to carry the wedge for part of its travel in a tilted position. Furthermore,

Figure 3-48. Bent Valve Stem with Damage to Stem Surface Resulting from Attempted Closure of Wedge Gate Valve with Wedge Rotated and Jammed in Valve Body, Size DN 300 (NPS 12) Class 150. (*Courtesy of David Mair.*)

the lengths of body rib and wedge guide were far too short to adequately support the wedge during all stages of travel.

This valve failure was not isolated but was typical for a high percentage of all installed wedge gate valves. Finally, all suspect valves had to be replaced prior to start-up of the plant.

Standards Pertaining to Wedge Gate Valves

Appendix C provides a list of U.S. and British standards pertaining to wedge gate valves.

Applications

Duty:
   Stopping and starting flow
   Infrequent operation
Service:
   Gases
   Liquids
   Rubber-seated wedge gate valves without bottom cavity for
      fluids carrying solids in suspension

Vacuum
Cryogenic

PLUG VALVES

Plug valves are rotary valves in which a plug-shaped closure member is rotated through increments of 90° to engage or disengage a porthole or holes in the plug with the ports in the valve body. The shape of the plug may thereby be cylindrical, as in the valves shown in Figure 3-49 through Figure 3-53, or tapered, as in the valves shown in Figure 3-54 through Figure 3-58. Rotary valves with ball-shaped plugs are likewise members of the plug valve family, but are conventionally referred to as ball valves. These valves are discussed separately on page 108.

The shape of the port is commonly rectangular in parallel plugs, and truncated triangular shapes in taper plugs. These shapes permit a slimmer valve construction of reduced weight, but at the expense of some pressure drop. Full area round-bore ports are normally used only if the pipeline has to be scraped or the nature of the fluid demands a full area round bore. However, some plug valves are made only with round-bore because of the method of sealing employed.

Figure 3-49. Lubricated Cylindrical Plug Valve. (*Courtesy of Pegler Hattersley Limited.*)

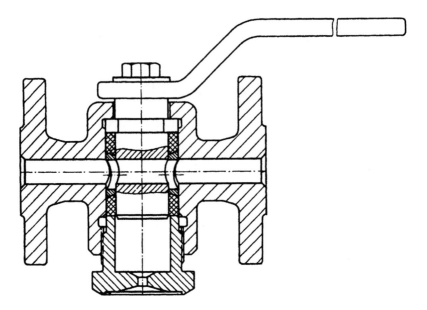

Figure 3-50. Cylindrical Plug Valve with Packing Sleeve. (*Courtesy of Rich. Klinger AG.*)

Plug valves are best suited for stopping and starting flow and flow diversion, though they are also used occasionally for moderate throttling, depending on the nature of the service and the erosion-resistance of the seatings. Because the seatings move against each other with a wiping motion, and in the fully open position are also fully protected from contact with the flowing fluid, plug valves are generally capable of handling fluids with solids in suspension.

Cylindrical Plug Valves

The use to which plug valves can be put depends to some extent on the way the seal between the plug and the valve body is produced. In the case of cylindrical plug valves, four sealing methods are frequently employed: by a sealing compound, by expanding the plug, by O-rings, and by wedging an eccentrically shaped plug into the seat.

The cylindrical plug valve shown in Figure 3-49 is a lubricated plug valve in which the seat seal depends on the presence of a sealing compound between the plug and the valve body. The sealing compound is introduced to the seatings through the shank of the plug by a screw or an injection gun.

Figure 3-51. Cylindrical Plug Valve with Expandable Split Plug. (*Courtesy of Langley Alloys Limited.*)

Thus, it is possible to restore a defective seat seal by injecting an additional amount of sealing compound while the valve is in service.

Because the seating surfaces are protected in the fully open position from contact with the flowing fluid, and a damaged seat seal can easily be restored, lubricated plug valves have been found to be particularly suitable for abrasive fluids. However, lubricated plug valves are not intended for throttling, although they are sometimes used for this purpose. Because throttling removes the sealing compound from the exposed seating surfaces, the seat seal must be restored, in this case, each time the valve is closed.

Unfortunately, the manual maintenance of the sealing compound is often a human problem. Automatic injection can overcome this problem, but it adds to the cost of installation. When the plug has become immovable in the valve body due to lack of maintenance or improper selection of the sealing compound, or because crystallization has occurred between the seatings, the valve must be cleaned or repaired.

The seat seal of the cylindrical plug valves shown in Figure 3-50 and Figure 3-51 depends on the ability of the plug to expand against the seat.

The plug of the valve shown in Figure 3-50 is fitted for this purpose with a packing sleeve, which is tightened against the seat by a follower nut.

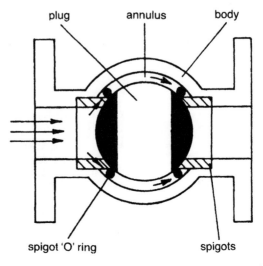

VALVE SHOWN IN CLOSED POSITION

Figure 3-52. Cylindrical Plug Valve with O-Ring Seat Seal. Valve Shown in Closed Position. (*Courtesy of Orseal Limited.*)

The packing commonly consists of compressed asbestos or solid PTFE. If the packing needs retightening to restore the seat tightness, this must be carried out while the valve is in the closed position to prevent the packing from expanding into the flow passage. The valve is made only in small sizes, but may be used for fairly high pressures and temperatures. Typical applications for the valve are the isolation of pressure gauges and level gauges.

The plug of the valve shown in Figure 3-51 is split into two halves and spread apart by a wedge, which may be adjusted from the outside. The seat seal is provided here by narrow PTFE rings that are inserted into the faces of the plug halves. By this method of sealing and seat loading, the valve is also capable of double block and bleed.

The valve is specifically intended for duties for which stainless steel and other expensive alloys are required, and it is capable of handling slurries, but not abrasive solids. Very tacky substances also tend to render the wedging mechanism inoperable. But within these limits, the valve has proved to be very reliable under conditions of frequent valve operation.

Figure 3-52 shows a parallel plug valve in which the seat seal is provided by O-rings. The O-rings are mounted on spigot-like projections of the upstream and downstream ports, which form the seats for the plug.

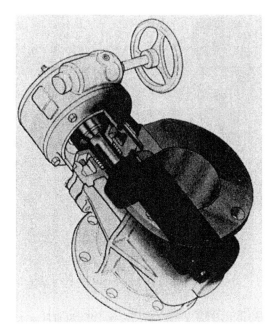

Figure 3-53. Cylindrical Plug Valve with Eccentric Semi-Plug. (*Courtesy of DeZurik.*)

When the valve is being closed, the fluid pressure enters the cavity between the plug and the valve body past the upstream O-ring and forces the downstream O-ring into intimate contact with the plug and the projections of the valve ports. The main application of this valve is for high-pressure hydraulic systems.

Efforts to eliminate most of the friction between the seatings and to control the sealing capacity of the valve by the applied torque led to the development of the eccentric plug valve shown in Figure 3-53 in which an eccentric plug matches a raised face in the valve body. By this method of seating, the valve is easy to operate and capable of handling sticky substances that are difficult to handle with other types of valves. The seating faces may be in metallic contact or lined with an elastomer or plastic in conjunction with the lining of the valve body.

Taper Plug Valves

Taper plug valves permit the leakage gap between the seatings to be adjusted by forcing the plug deeper into the seat. The plug may also be

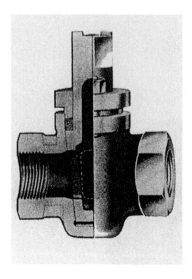

Figure 3-54. Taper Plug Valve with Unlubricated Metal Seatings. (*Courtesy of Pegler Hattersley Limited.*)

rotated while in intimate contact with the valve body, or lifted out of the valve body prior to rotation and seated again after being rotated through 90°, depending on design.

Figure 3-54 shows a taper plug valve with unlubricated metal seatings. Because the friction between the seatings is high in this case, the permissible seating load is limited to keep the plug freely movable. The leakage gap between the seatings is therefore relatively wide, so the valve achieves a satisfactory seat seal only with liquids that have a high surface tension or viscosity. However, if the plug has been coated with a grease prior to installation, the valve may be used also for wet gases such as wet and oily compressed air.

The lubricated taper plug valve shown in Figure 3-55 is similar to the lubricated cylindrical plug valve except for the shape of the plug. Both valves may also serve the same duties. However, the lubricated taper plug valve has one operational advantage. Should the plug become immovable after a prolonged static period or as a result of neglected lubrication, the injection of additional sealing compound can lift the plug off the seat just enough to allow the plug to be moved again. When freeing the plug in this way, the gland should not be slackened as is sometimes done by users, but should rather be judiciously tightened. On the debit side, it is also possible to manually overtighten the plug and cause the plug to seize in this way.

The valve shown in Figure 3-56 is a lubricated taper plug valve in which the plug is mounted in the inverted position and divorced from the stem.

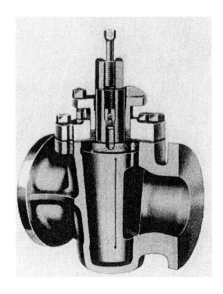

Figure 3-55. Lubricated Taper Plug Valve. (*Courtesy of Serck Audco Pty. Ltd.*)

The plug is adjusted in its position by a screw in the valve cover while the sealant is injected into the body at the stem end. To prevent the fluid pressure from driving the plug into the seat, the plug ends are provided with balance holes that permit the fluid pressure to enter the cavities at both plug ends. By this design the plug valve may be used for very high pressures without becoming inoperable because of the fluid pressure driving the plug into the seat.

Efforts to overcome the maintenance problem of lubricated plug valves led to the development of the taper plug valve shown in Figure 3-57 in which the plug moves in a PTFE body sleeve. The PTFE body sleeve keeps the plug from sticking, but the operating torque can still be relatively high due to the large seating area and the high seating stress. On the other hand, the large seating area gives good protection against leakage should some damage occur to the seating surface. As a result, the valve is rugged and tolerates abusive treatment. The PTFE sleeve also permits the valve to be made of exotic materials that would otherwise tend to bind in mutual contact. In addition, the valve is easily repaired in the field, and no lapping of the plug is required.

The taper plug valve shown in Figure 3-58 is designed to eliminate most of the sliding between the seatings. This is achieved by lifting the plug out of the body seat by means of a hinged level arrangement prior to it being rotated and reseating it after it is rotated to the desired position.

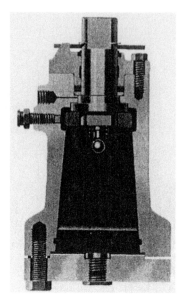

Figure 3-56. Lubricated Taper Plug Valve with Inverted Pressure-Balanced Plug. (*Courtesy of Nordstrom.*)

Figure 3-57. Taper Plug Valve with PTFE Body Sleeve. (*Courtesy of Xomox Corporation, "Tufline."*)

The plug may be rubber faced, and is, by its taper, normally self-locking. This particular valve is a multiport companion valve to the eccentric plug valve shown in Figure 3-53.

Antistatic Device

In plug valves, seats and packings made of a polymeric material such as PTFE can electrically insulate the plug and the valve stem from the valve

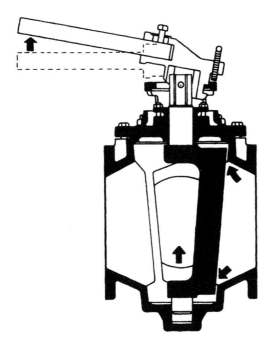

Figure 3-58. Multiport Taper Plug Valve with Lift Plug. (*Courtesy of DeZurik.*)

body. Under these conditions, friction from the flowing fluid may generate an electrostatic charge in the plug and the stem that is high enough to produce an incendiary spark. This possibility is more pronounced with two-phase flow. If the fluid handled by the valve is flammable, the valve must be provided with an antistatic device, which achieves electrical continuity between the plug, stem, and the valve body.

Plug Valves for Fire Exposure

Plug valves, which may be exposed to a plant fire when handling a flammable fluid, must remain essentially fluid-tight internally and externally and be operable during and after a fire. This degree of resistance to fire damage is particularly difficult to achieve when the plug valve normally relies on polymers for seat and stem seal. Common practice in this case is to provide the valve with an auxiliary metal seat in close proximity to the plug, against which the plug can float after the soft seat has disintegrated. The soft stem packing can readily be replaced with a fire-resistant packing.

The requirements for testing and evaluating the performance of valves exposed to fire are similar to those for ball valves, described on pages 108 and 109.

Multiport Configuration

Plug valves adapt readily to multiport configurations such as those shown in Figure 3-59. The valves may be designed for transflow, in which case the second flow passage opens before the first closes; or for non-transflow, in which case the first flow passages closes before the second opens. The transflow sequence is intended for duties in which the flow cannot be momentarily interrupted; for example, on the outlet of a positive displacement pump that is not protected by a relief valve. The non-transflow sequence may be required when a momentary flow into one port from the other is not permissible; for example, at the outlet of a measuring vessel; but the plug is not normally intended to shut off fluid-tight in the intermediate position. In most practical applications, however, when valves are operated fairly quickly, there is little difference between the effects of transflow and non-transflow in fluid flow.

The direction of flow through lubricated multiport valves should be such that the fluid pressure forces the plug against the port that is to be shut off.

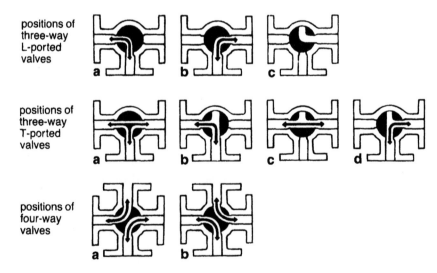

Figure 3-59. Multiport Configuration of Plug Valves. (*Courtesy of Serck Audco Pty. Ltd.*)

If the pressure acts from the opposite direction, lubricated plug valves will not hold their rated pressure.

Face-to-Face Dimensions and Valve Patterns

The designers of the U.S. and British standards of cast iron and carbon steel plug valves have attempted to make the face-to-face dimensions of plug valves identical to those of gate valves. To accommodate plug valves within these dimensions, some concessions had to be made on the flow area through the plug in the lower pressure ratings. But even with these concessions, plug valves for class 150 can be made interchangeable with gate valves up to DN 300 (NPS 12) only. This limitation led to the introduction of an additional long-series plug valve for class 150. As a result, the following valve patterns have emerged:

1. Short pattern, having reduced area plug ports and face-to-face dimensions that are interchangeable with gate valves. This pattern applies to sizes up to DN 300 (NPS 12) in class 150 only.
2. Regular pattern, having plug ports with areas larger than short or venturi patterns. The face-to-face dimensions are interchangeable with gate valves for pressure ratings class 300 and higher. Regular pattern plug valves for class 150 have face-to-face dimensions to a long series that are not interchangeable with gate valves.
3. Venturi pattern, having reduced area ports and body throats that are approximately venturi. The face-to-face dimensions are interchangeable with gate valves for pressure ratings class 300 and higher. Venturi-pattern plug valves for class 150 have face-to-face dimensions to a long series that are not interchangeable with gate valves.
4. Round-port full-area pattern, having full-area round ports through the valve. The face-to-face dimensions are longer than short, regular, or venturi-pattern plug valves and are noninterchangeable with gate valves.

Standards Pertaining to Plug Valves

Appendix C provides a list of U.S. and British standards pertaining to plug valves.

Applications

Duty:
    Stopping and starting flow
    Moderate throttling
    Flow diversion
Fluids:
    Gases
    Liquids
    Non-abrasive slurries
    Abrasive slurries for lubricated plug valves
    Sticky fluids for eccentric and lift plug valves
    Sanitary handling of pharmaceuticals and food stuffs
    Vacuum

BALL VALVES

Ball valves are a species of plug valves having a ball-shaped closure member. The seat matching the ball is circular so that the seating stress is circumferentially uniform. Most ball valves are also equipped with soft seats that conform readily to the surface of the ball. Thus, from the point of sealing, the concept of the ball valve is excellent. The valves shown in Figure 3-60 through Figure 3-66 are typical of the ball valves available.

The flow-control characteristic that arises from a round port moving across a circular seat and from the double pressure drop across the two seats is very good. However, if the valve is left partially open for an extended period under conditions of a high pressure drop across the ball, the soft seat will tend to flow around the edge of the ball orifice and possibly lock the ball in that position. Ball valves for manual control are therefore best suited for stopping and starting flow and moderate throttling. If flow control is automatic, the ball is continuously on the move, thus keeping this failure from normally occurring.

Because the ball moves across the seats with a wiping motion, ball valves will handle fluids with solids in suspension. However, abrasive solids will damage the seats and the ball surface. Long, tough fibrous material may also present a problem, as the fibers tend to wrap around the ball.

To economize in the valve construction, most ball valves have a reduced bore with a venturi-shaped flow passage of about three-quarters the nominal valve size. The pressure drop across the reduced-bore ball valve is thereby so small that the cost of a full-bore ball valve is not normally justified. However, there are applications when a full-bore ball valve is required, as for example, when the pipeline has to be scraped.

Seat Materials for Ball Valves

The most important seat material for ball valves is PTFE, which is inert to almost all chemicals. This property is combined with a low coefficient of friction, a wide range of temperature application, and excellent sealing properties. However, the physical properties of PTFE include also a high coefficient of expansion, susceptibility to cold flow, and poor heat transfer. The seat must therefore be designed around these properties. Plastic materials for ball valve seats also include filled PTFE, nylon, and many others. However, as the seating material becomes harder, the sealing reliability tends to suffer, particularly at low-pressure differentials. Elastomers such as buna-N are also used for the seats, but they impose restrictions on fluid compatibility and range of temperature application. In addition, elastomers tend to grip the ball, unless the fluid has sufficient lubricity. For services unsuitable for soft seatings, metal and ceramic seatings are being used.

Seating Designs

The intimate contact between the seatings of ball valves may be achieved in a number of ways. Some of the ones more frequently used are:

1. By the fluid pressure forcing a floating ball against the seat, as in the valves shown in Figure 3-60 through Figure 3-63.
2. By the fluid pressure forcing a floating seat ring against a trunnion-supported ball, as in the valve shown in Figure 3-64.
3. By relying mainly on the installed prestress between the seats and a trunnion-supported ball, as in the valve shown in Figure 3-65.
4. By means of a mechanical force, which is introduced to the ball and seat on closing, as in the valve shown in Figure 3-66.

Ball valves are also available in which the seal between the seat and ball is achieved by means of a squeeze ring such as an O-ring.

The first sealing method, in which the seating load is regulated by the fluid pressure acting on the ball, is the most common one. The permissible operating pressure is limited in this case by the ability of the downstream seat ring to withstand the fluid loading at the operating temperature without permanent gross deformation.

The seat rings of the valves shown in Figure 3-60 and Figure 3-61 are provided with a cantilevered lip, which is designed so that the ball contacts initially only the tip of the lip. As the upstream and downstream seats are pre-stressed on assembly against the ball, the lips deflect and put the seat rings into torsion. When the valve is being closed against the line pressure, the lip of the downstream seat deflects still further until finally the entire seat surface matches the ball. By this design, the seats have some spring action that promotes good sealing action also at low fluid pressures. Furthermore, the resilient construction keeps the seats from being crushed at high fluid loads.

The seat rings of the valve shown in Figure 3-60 are provided with peripheral slots, which are known as pressure-equalizing slots. These slots reduce the effect of the upstream pressure on the total valve torque. This is achieved by letting the upstream pressure filter by the upstream seat ring into the valve body cavity so that the upstream seat ring becomes pressure balanced.

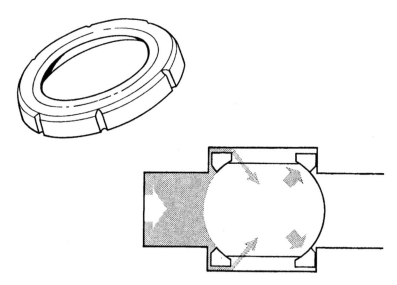

Figure 3-60. Schematic View of Ball Valve with Floating Ball and Torsion Seats, Showing Function of Pressure-Equalizing Slots in Periphery of Seats. (*Courtesy of Worcester Valve Co., Ltd.*)

Figure 3-61. Ball Valve with Floating Ball and Torsion Seats, with Axial-Entry Body. (*Courtesy of Jamesbury International Corp.*)

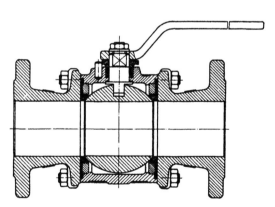

Figure 3-62. Ball Valve with Floating Ball and Diaphragm-Supported Seats, with Sandwich-Split Body. (*Courtesy of Rich. Klinger AG.*)

The valves shown in Figure 3-62 and Figure 3-63 are also designed to ensure a preload between the seatings. This is achieved in the valve shown in Figure 3-62 by supporting the seat rings on metal diaphragms that act as springs on the back of the seat rings. The seating preload of the valve shown in Figure 3-63 is maintained by a spring that forces the ball and

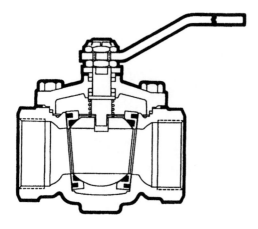

Figure 3-63. Ball Valve with Wedge Seats Spring-Loaded from the Top, with One-Piece Top-Entry Body. (*Courtesy of McCanna*)

Figure 3-64. Ball Valve with Trunnion-Supported Ball and Floating Seats, with One-Piece Sealed Body. (*Courtesy of Cameron Iron Works, Inc.*)

seat-ring assembly from the top into wedge-shaped seat ring back faces in the valve body.

The design of the ball valve shown in Figure 3-64 is based on the second seating method in which the fluid pressure forces the seat ring against a trunnion-supported ball. The floating seat ring is sealed thereby peripherally by an O-ring. Because the pressure-uncompensated area of the seat ring can be kept small, the seating load for a given pressure rating can be regulated to suit the bearing capacity of the seat. These valves may

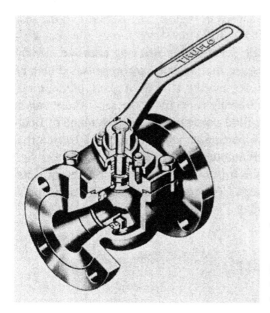

Figure 3-65. Ball Valve with Trunnion-Supported Ball and Wedge Seats Forced into the Valve Body by the Bonnet, with One-Piece Top-Entry Body. (*Courtesy of Truflo Limited.*)

therefore be used for high fluid pressures outside the range of floating-ball type ball valves. This particular valve also has a device that rotates the seat rings by a small amount each time the valve is operated. The purpose of this rotating action is to evenly distribute the seat wear. Should the seat seal fail, a temporary seat seal can be provided by the injection of a sealant to the seatings.

The third seating method in which the seat seal depends mainly on the installed prestress between the seats and a trunnion-supported ball, as in the valve shown in Figure 3-65, is designed to limit the operating torque of the valve at high fluid pressures. The lips around the ports of the ball are radiused (rounded) to reduce the seating interference when the ball is in the open position. When the ball is moved into the closed position, the seating interference increases. If the valve is required for double block and bleed, the back of the seat rings must be provided with an elastomeric O-ring.

The fourth seating method in which the seating load is regulated on closing by an introduced mechanical force is designed to avoid most of the sliding action between the seatings. In the valve shown in Figure 3-66, this is achieved by a cam mechanism that lifts the ball out of the seat prior to opening the valve and forces the ball back into the seat after closing the valve.

Figure 3-66. Ball Valve with Cam Mechanism to Seat and Unseat Ball. (*Courtesy of Orbit Valve Company.*)

Pressure-Equalizing Connection

Double-seated ball valves may contain a sealed valve body cavity in both the open and closed valve positions. When the valve is closed, the sealed cavity extends between the upstream and downstream seats. When the valve is open, a sealed cavity may exist also between the ball and the valve body. If these cavities are filled with a liquid of high thermal expansion, the pressure rise in these cavities may overstress some valve components because of thermal expansion of the trapped fluid, unless the excess fluid pressure can be relieved.

The cavity between the ball and the valve body is normally relieved to the flow passage via a hole in the top or bottom flank of the ball. If the valve is closed, the excess pressure in the cavity between the seats may be relieved in various ways.

In the case of ball valves with floating seats, as shown in Figure 3-64, excess pressure in the valve body will open the upstream seat seal where the least pressure differential exists. This permits the excess pressure to escape.

In other double-seated ball valves, however, the fluid pressure must overcome the prestress between the ball and the upstream seat. If the seat rings are provided with some springing action, as in the valves shown in Figure 3-60 through Figure 3-62, the fluid pressure may be able to open the upstream seat seal without becoming excessively high. On the other hand, if the seat rings are of a more rigid construction, thermal expansion of the trapped fluid may create an excessively high pressure in the sealed cavity, depending on the prestress between the upstream seatings. In this case, the upstream flank of the ball is usually provided with a pressure-equalizing hole, thus permitting flow through the valve in one direction only. If the valve catalog does not advise on the need for pressure-equalizing connection, the manufacturer should be consulted. The provision of a pressure-equalizing connection is not normally standard with ball valves except for cryogenic service. The pressure-equalizing connection is necessary in that case because of the rigidity of normally soft plastics at low temperatures, which tends to resist the opening of the upstream seat seal.

Antistatic Device

The polymeric seats and packings used in ball valves can electrically insulate the ball and the stem from the valve body. Ball valves may therefore have to be provided with an antistatic device for the same reason as described on page 98 for plug valves. Figure 3-67 shows a typical antistatic device consisting of spring-loaded plungers—one fitted between the tongue of the stem and the ball, and a second between the stem and the body.

Ball Valves for Fire Exposure

The soft seals for seat and stem commonly used in ball valves will disintegrate if the valve is exposed to fire for a long enough period. If such valves are used for flammable fluids, they must be designed so that loss of the soft seals due to an external fire does not result in gross internal and external valve leakage. Such designs provide emergency seals for seat and stem that come into operation after the primary seals have failed.

The emergency seat seal may be provided by a sharp-edged or chamfered secondary metal seat in close proximity to the ball, so that the ball can float against the metal seat after the soft seating rings have disintegrated.

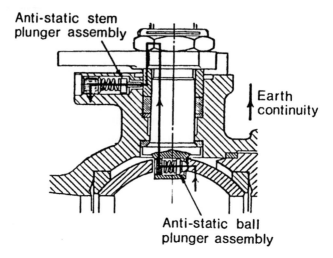

Anti-static stem
plunger assembly

Earth
continuity

Anti-static ball
plunger assembly

Figure 3-67. Antistatic Device for Grounding Stem to Ball and Stem to Body. (*Courtesy of Worcester Valve Co., Ltd.*)

The stuffing box may be fitted with an auxiliary asbestos-based or pure graphite packing, or the packing may be made entirely of an asbestos compound or pure graphite.

Numerous standards have been established that cover the requirements for testing and evaluating the performance of soft-seated ball valves when exposed to fire. The three basic standards are BS 5146, API 607, and API RP 6F.

BS 5146 is a derivative of OCMA FA VI, which itself was taken from a test specification created by Esso Petroleum in the U.K. This test differed from all former ones by requiring the valve to be in the open position during the test and by using a flammable liquid in the valve. The test owes its origin to the recognition by Esso that, in an actual plant fire, a significant number of valves may be in the open position and must be subsequently closed. Moves are under foot by a number of standard organizations to arrive at an international fire-test specification.

Besides paying attention to the fire testing of ball valves in systems handling flammable fluids, similar attention must be paid to the effect of fire on the entire fluid handling system including valves other than ball valves, valve operators, pumps, filters, pressure vessels, and, not least, the pipe flanges, bolting, and gaskets.

Fire-tested ball valves are referred to as fire-safe. However, this term is unacceptable to valve manufacturers from the product liability standpoint.

Multiport Configuration

Ball valves adapt to multiport configurations in a manner similar to plug valves, previously discussed on page 98.

Ball Valves for Cryogenic Service

Ball valves are used extensively in cryogenic services, but their design must be adapted for this duty. A main consideration in the design of these valves is the coefficient of thermal contraction of the seat ring material, which is normally higher than that of the stainless steel of the ball and valve body. The seat rings shrink, therefore, on the ball at low temperatures and cause the operating torque to increase. In severe cases, the seat ring may be overstressed, causing it to split.

This effect of differential thermal contraction between the seats and the ball may be combated by reducing the installed prestress between the seats and the ball by an amount that ensures a correct prestress at the cryogenic operating temperature. However, the sealing capacity of these valves may not be satisfactory at low fluid pressures if these valves also have to operate at ambient temperatures.

Other means of combating the effect of differential thermal contraction between the seats and the ball include supporting the seats on flexible metal diaphragms; choosing a seat-ring material that has a considerably lower coefficient of contraction than virgin PTFE, such as graphite or carbon filled PTFE; or making the seat rings of stainless steel with PTFE inserts in which the PTFE contents are kept to a minimum.

Because plastic seat-ring materials become rigid at cryogenic temperatures, the surface finish of the seatings and the sphericity of the ball must be of a high standard to ensure a high degree of seat tightness. Also, as with other types of valves for cryogenic service, the extended bonnet should be positioned no more than 45° from the upright to ensure an effective stem seal.

Variations of Body Construction

Access to the ball valve internals can be provided in various ways. This has led to the development of a number of variations in the body construction; Figure 3-68 shows the most common variations.

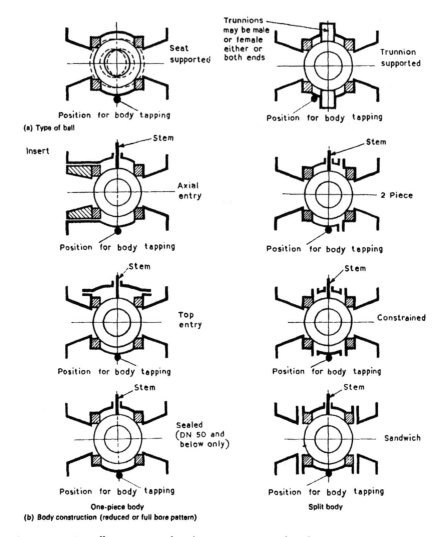

Figure 3-68. Different Types of Body Construction and Body Tapping Locations. (*Reprinted from BS 5159:1974, Courtesy of British Standards Institution.*)

The one-piece body has the fewest number of body joints subject to pipeline stresses. This type of body therefore is often chosen for hazardous fluids. If the valve is to be buried, the sealed-body variety is frequently used. The one-piece body with top entry and the various split body constructions offer easy entry to the valve internals. In the case of welded-in-line valves, those with top entry may also be serviced *in situ*. A selection from these types is often a matter of personal preference.

Face-to-Face Dimensions

The original practice of U.S. manufacturers was to make the face-to-face dimensions of flanged ball valves to the nearest valve standard, which gave minimum material content. This happened to be the gate valve standard, but the face-to-face dimensions of class 150 and of sizes DN 200 (NPS 8) through DN 300 (NPS 12) of class 300 permitted only reduced-bore construction.

In 1961, when UK manufacturers also introduced the flanged ball valve, there was an additional demand for full-bore ball valves. When it was impossible to accommodate the full-bore ball valve in the confines of the face-to-face dimensions of gate valves, the face-to-face dimensions of regular-pattern plug valves were adopted.

Thus, there is a short and a long series of ball valves for class 150, and in sizes DN 200 (NPS 8) through DN 300 (NPS 12), and for class 300—one for reduced-bore and one for full-bore ball valves, respectively. In the case of the higher-pressure ratings, the face-to-face dimensions of gate valves accommodate both reduced-bore and full-bore ball valves throughout.

The master standard for face-to-face dimensions is ISO 5752. This standard includes all the recognized dimensions worldwide that are used in the piping industry. However, ISO 5752 does not try to define reduced-bore or full-bore except for sizes DN 200 (NPS 8) through DN 300 (NPS 12) of class 300.

Standards Pertaining to Ball Valves

Appendix C provides a list of U.S. and British standards pertaining to ball valves.

Applications

Duty:
  Stopping and starting flow
  Moderate throttling
  Flow diversion
Service:
  Gases
  Liquids

Non-abrasive slurries
Vacuum
Cryogenic

BUTTERFLY VALVES

Butterfly valves are rotary valves in which a disc-shaped closure member is rotated through 90°, or approximately, to open or close the flow passage. The butterfly valves shown in Figure 3-69 through Figure 3-83 represent a cross section of the many variations available.

The original butterfly valve is the simple pipeline damper that is not intended for tight shut-off. This valve is still an important member of the butterfly valve family.

The advent of elastomers has initiated the rapid development of tight shut-off butterfly valves in which the elastomer serves as the sealing element between the rim of the disc and the valve body. The original use of these valves was for water.

As more chemical-resistant elastomers became available, the use of butterfly valves spread to wide areas of the process industries. The elastomers used for these purposes must not only be corrosion-resistant but also abrasion-resistant, stable in size, and resiliency-retentive, that is, they must not harden. If one of these properties is missing, the elastomer may be unsuitable. Valve manufacturers can advise on the selection and limitations of elastomers for a given application.

Efforts to overcome some of the limitations of elastomers led to the development of butterfly valves with PTFE seats. Other efforts led to the development of tight shut-off butterfly valves with metal seatings. By these developments, butterfly valves became available for a wide range of pressures and temperatures, based on a variety of sealing principles.

Butterfly valves give little resistance to flow when fully open and **sensitive flow control** when open between about 15° and 70°. Severe throttling of liquids may, of course, lead to cavitation, depending on the vapor pressure of the liquid and the downstream pressure. Any tendency of the liquid to cavitate as a result of throttling may be combated partly by sizing the butterfly valve smaller than the pipeline so that throttling occurs in the near half-open position, and/or by letting the pressure drop occur in steps, using a number of valves, as discussed in Chapter 2. Also, if the butterfly valve is closed too fast in liquid service, waterhammer may become excessive.

By closing the butterfly valve slowly, excessive waterhammer can be avoided, as discussed in Chapter 2.

Because the disc of butterfly valves moves into the seat with a wiping motion, most butterfly valves are capable of handling fluids with solids in suspension and, depending on the robustness of the seatings, also powders and granules. In horizontal pipelines, butterfly valves should be mounted with the stem in the horizontal. Furthermore, when opening the valve, the bottom of the disc should lift away from solids that may have accumulated on the upstream side of the disc.

Seating Designs

From the point of seat tightness, butterfly valves may be divided into nominal-leakage valves, low-leakage valves, and tight shut-off valves. The nominal- and low-leakage valves are used mainly for throttling or flow control duty, while tight shut-off butterfly valves may be used for tight shut-off, throttling, or flow control duty.

The butterfly valves shown in Figure 3-69 are examples of nominal- and low-leakage butterfly valves in which both the seat and disc are metallic. For applications in which a lower leakage rate is required, butterfly valves that have a piston ring on the rim of the disc are available.

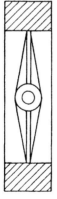 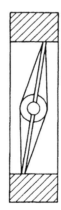 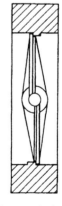

swing-thru disc and metal seating, not fluid tight

angle-seated disc and metal seating, not fluid tight

step-seated disc and metal seating, not fluid tight

Figure 3-69. Butterfly Valves. (*Courtesy of GEC-Elliot Control Valves Limited.*)

Figure 3-70. Butterfly Valve with Resilient Replaceable Liner and Interference-Seated Disc. (*Courtesy of Keystone International, Inc.*)

The intimate contact between the seatings of tight shut-off butterfly valves may be achieved by various means. Some of the ones more frequently used are:

1. By interference seating that requires the disc to be jammed into the seat.
2. By forcing the disc against the seat, requiring the disc to tilt about a double offset hinge in a manner comparable to tilting disc check valves, as described on page 154.
3. By pressure-energized sealing using sealing elements such as O-rings, lip seals, diaphragms, and inflatable hoses.

The majority of butterfly valves are of the interference-seated type in which the seat has a rubber liner, as in the valves shown in Figure 3-70 through Figure 3-72. Where rubber is incompatible with the fluid to be sealed, the liner may be made of PTFE, which is backed-up by an elastomer cushion to impart resiliency to the seat.

There are a number of rubber liner constructions in common use. Typical constructions are:

1. Rubber liner consisting of a U-shaped ring that is slipped over the body without bonding, as in the valve shown in Figure 3-70. Such seats are readily replaceable. If the liner is made of a relatively rigid

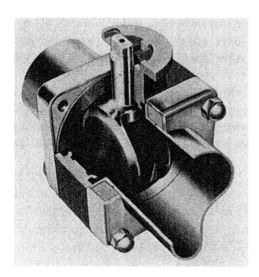

Figure 3-71. Butterfly Valve with Resilient Liner and Interference-Seated Double Disc for Double Block and Bleed. Used for Isolating Food Stuffs from Cleaning-in-Place Cleaning Fluids. (*Courtesy of Amri SA.*)

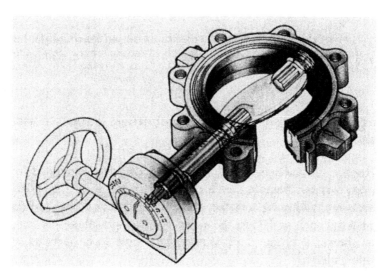

Figure 3-72. Butterfly Valve with Resilient Replaceable Liner and Interference-Seated Disc, Liner Bonded to Steel Band, Valve Body Split in Two Halves. (*Courtesy of DeZurik.*)

material such as PTFE, the valve body is split along the centerline as shown in Figure 3-72 to permit the liner to be inserted without manipulation.

2. Rubber liner that is bonded to the valve body. This construction minimizes the wearing effects of rubber bunching from disc rotation that

may occur if the liner is loose. On the debit side, the rubber liner cannot be replaced.
3. U-shaped rubber liner that is bonded to a metal band. This combination can be inserted into a split valve body, as in the valve shown in Figure 3-72.

To ensure that rubber-lined butterfly valves achieve their full sealing capacity, they must be correctly installed. Two requirements must be observed:

1. The rubber liner should be fully supported by the pipe flanges. The pipe flanges should therefore be of the weld-neck type rather than the slip-on type. In the case of slip-on flanges, the rubber liner remains unsupported between the valve bore and the outside diameter of the pipe. This lack of support of the liner tends to promote distortion of the liner during valve operation, resulting in early wear and seat leakage.
2. When installing the valve, the disc must initially be put into the near-closed position so as to protect the rim of the disc from damage during handling. Prior to tightening the flange bolts, the disc must be rotated into the fully open position to permit the liner to find its undisturbed position.

Figure 3-73 illustrates the precautions that must be taken when installing rubber-lined butterfly valves.

The flange of the rubber liner also serves as a sealing element against the pipeline flanges. The installation of additional rubber gaskets between pipe flanges and valve would tend to reduce the support of the rubber liner and, consequently, reduce the sealing capacity of the valve.

The scrap view shown in Figure 3-74 belongs to an interference-seated butterfly valve that carries the sealing element on the rim of the disc. The sealing element consists in this case of a heavy section O-ring with a tail clamped to the disc. By adjusting the clamping force, the seating interference can be adjusted, within limits. Because the sealing element deforms against a wide face instead of around the narrow face of a disc, the seating and unseating torques are correspondingly lower. This particular make of valve is made in larger sizes only and is used for relatively high fluid pressures.

The butterfly valves shown in Figure 3-75 and Figure 3-76 rely for a seat seal on pressure energized rubber elements.

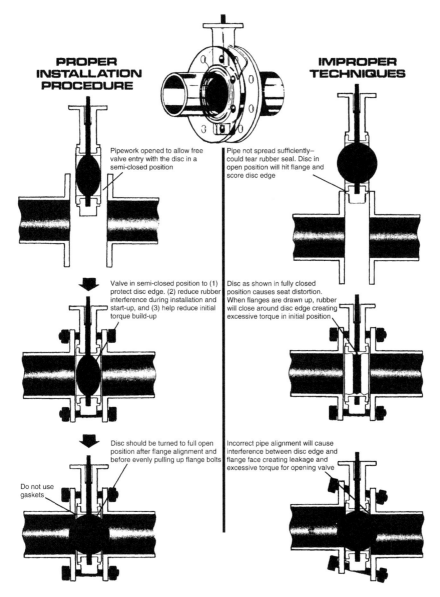

Figure 3-73. Proper and Improper Installation Procedures for Interference-Seated Butterfly Valves with Replaceable Rubber Liner. (*Courtesy of Keystone International, Inc.*)

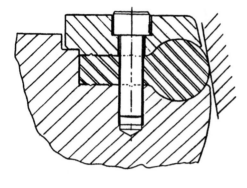

Figure 3-74. Scrap View of Interference-Seated Butterfly Valve Showing the Resilient Sealing Element Carried on the Rim of the Disc. (*Courtesy of Boving & Co., Limited.*)

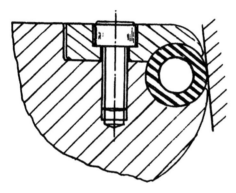

Figure 3-75. Scrap View of Butterfly Valve showing an Inflatable Sealing Element Carried on the Rim of the Disc. (*Courtesy of Boving & Co., Limited.*)

The sealing element of the seating arrangement shown in Figure 3-75 consists of an inflatable hose mounted on the rim of the disc. The hose is reinforced inside by a metallic conduit and connected through the operating shaft to the upstream system or an external fluid pressure system. The disc is moved into the tapered seat with the hose deflated so that the seating torque is minimal. The hose is then pressurized to provide a fluid-tight seal against the seat. If the seal requires further tightening, the hose may be pumped up using a hand pump. When the valve is to be opened, the hose is first deflated so that the valve opens with a minimum of unseating torque. The valve is made to the largest sizes in use.

The sealing element of the butterfly valve shown in Figure 3-76 consists of a tubular shaped diaphragm of T-cross section, which is mounted in a slot of the valve body and sealed against the flow passage. The diaphragm is pressurized on closing against the rim of the disc and depressurized on opening in a manner similar to the valve shown in Figure 3-75.

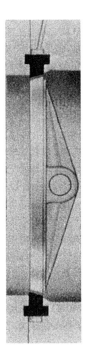 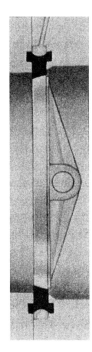

Figure 3-76. Butterfly Valve with Inflatable Sealing Element Carried in a Recess of the Valve Body. (*Courtesy of GEC-Elliot Control Valves, Limited.*)

Efforts to adapt butterfly valves to wider temperature and pressure ranges have led to the development of a family of butterfly valves that may be fitted with seatings of a variety of construction materials to meet the operational requirements. Such seatings may be metal-to-polymer or metal-to-metal, and may be designed to satisfy the requirements of fire-tested valves. The majority of these valves may be used for flow in both directions.

Valves of this performance class have acquired the name high-performance butterfly valves. The name is taken to mean that this type of butterfly valve has a greater pressure-temperature envelope than the common elastomer-lined or seated butterfly valves. Figure 3-77 through Figure 3-83 show examples of such valves. However, the illustrations do not show all the seal variations available for each particular valve.

Figure 3-77 illustrates a high-performance butterfly valve in which the seat seal is achieved by forcing the disc against the seat about a double offset hinge in a manner comparable to tilting disc check valves. The shape of the disc represents a slice from a tapered plug made at an oblique angle to the plug axis. The seating faces formed in this way are tapered on an elliptical circumference. The disc rotates around a point below the centerline of the

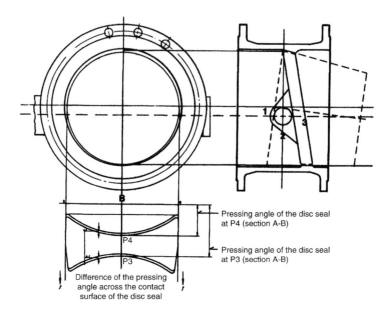

Pressing angle of the disc seal
at P4 (section A-B)

Pressing angle of the disc seal
at P3 (section A-B)

Difference of the pressing
angle across the contact
surface of the disc seal

Figure 3-77. Metal-Seated High-Performance Butterfly Valve with Tapered Seating Faces and Sealing of the Disc at a Non-Interlocking Angle. (*Courtesy of Clow Corporation.*)

valve and behind the seat face so that the disc drops into and lifts out of the seat with little rubbing action. By this concept, the seating load is provided by the applied closing torque from the valve operator and the hydrostatic torque from the fluid pressure acting on the unbalanced area of the disc. The valve may be provided with metal-to-metal seatings as illustrated, or the disc may be provided with a variety of soft sealing elements that seal against the metal seat provided by the body. With this choice of seating constructions, the valve may be used for differential pressures up to 35 bar (500 lb/in.^{2}) and operating temperatures between $-200°C$ ($-350°F$) and $650°C$ ($1200°F$).

The concept of the high-performance butterfly valve shown in Figure 3-78 may be likened to that of a ball valve that uses a wafer section. The illustrated seat consists of a U-shaped plastic seal ring that is mounted in a T-slot of the valve body and backed up by an elastomeric O-ring. The O-ring imparts some initial compression stress to the seat. As the disc moves into the seat with slight interference, fluid pressure acting on the O-ring forces the seat ring into closer contact with the rim of the disc.

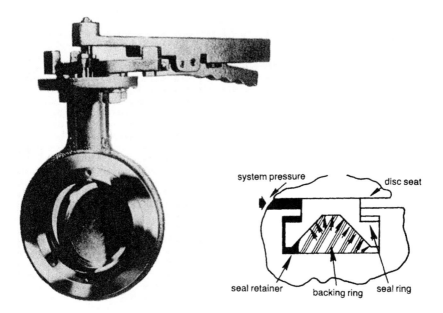

Figure 3-78. High-Performance Butterfly Valve with O-Ring-Backed Plastic Seat. (*Courtesy of Posi-Seal International, Inc.*)

The valve is made with a variety of seat ring constructions, including metal construction to suit a variety of operating conditions.

The concept of the high-performance butterfly valve shown in Figure 3-79 is also borrowed from the ball valve, but with a difference from the valve shown in Figure 3-78 that the axis of the disc is offset not only from the plane of the seat, but also by a small amount from the centerline of the valve. In this way, the disc moves into the seat in a camming action, thereby moving progressively into intimate contact with the seat. Conversely, the seatings rapidly disengage during opening so that there is minimum rubbing between the seatings. The seat, as illustrated, consists of a PTFE lip seal with tail, which is clamped between the valve body and a retainer. Once the disc has entered the seat with slight interference, fluid pressure forces the lip of the seat into more intimate contact with the rim of the disc, irrespective of whether the seat is located on the upstream or downstream side of the disc. This ability of the seatings to seal in both directions is aided by a small but controlled amount of axial movement of the disc. The valve is also available with metal and fire-tested seat constructions.

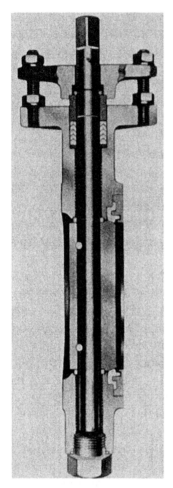

Figure 3-79. High-Performance Butterfly Valve with Flexible PTFE Lip Seat. (*Courtesy of Jamesbury International Corp.*)

The principle of the double offset location of the disc axis in combination with pressure-energized sealing in both directions applies also to the high-performance butterfly valves shown in Figure 3-80 through Figure 3-82. The high-performance butterfly valve shown in Figure 3-80 contains, like the valve shown in Figure 3-79, a PTFE lip seal for the seat, but of a modified shape in which the lip is supported by a titanium ring. The seat is also available in metal and fire-tested construction.

The high-performance butterfly valve shown in Figure 3-81 uses an elastomeric O-ring for the seat that is encapsulated in PTFE and anchored in the valve body. The O-ring imparts elasticity and resiliency to the seat while the PTFE envelope protects the O-ring from the effects of the system fluid.

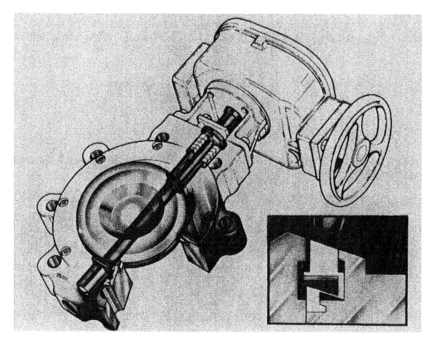

Figure 3-80. High-Performance Butterfly Valve with PTFE Lip Seal Supported by Titanium Metal Ring. (*Courtesy of DeZurik.*)

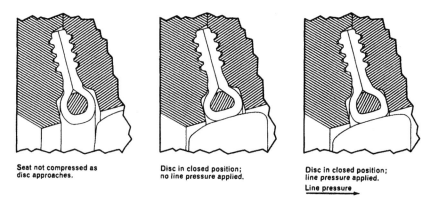

Seat not compressed as disc approaches.

Disc in closed position; no line pressure applied.

Disc in closed position; line pressure applied.

Line pressure →

Figure 3-81. Scrap View of High-Performance Butterfly Valve Showing PTFE Encapsulated Elastomeric O-Ring Seat. (*Courtesy of Bray Valve and Controls.*)

PTFE seat incorporating
pliant membrane

Metal seat with PTFE insert

Figure 3-82. High-Performance Butterfly Valve. (*Courtesy of Xomox Corporation.*)

The seat is also available in fire-tested construction. The high-performance butterfly valve shown in Figure 3-82 incorporates a seat that is combined with a flexible membrane designed to return the seat to its original position each time the valve has been operated. One seat version consists of a T-shaped PTFE member that embeds a pliant membrane. A second fire-tested version consists of a U-shaped metal member with a PTFE insert that is carried on a metal membrane. The PTFE insert provides the seat seal under normal operating conditions. If the PTFE insert is destroyed in a fire, the metal seat takes over the sealing function.

The high-performance butterfly valve shown in Figure 3-83 differs from the previously described high-performance butterfly valve in that it relies on interference seating for seat tightness, using a double offset disc. The seat is available either in plastic, elastomer, metal, or plastic-metal fire-tested construction to suit the duty for which the valve is to be used. The performance of the plastic and metal seats relies on a back-up wire winding that is designed to provide absolute seat rigidity when the valve is closed but to avoid imposing a load on the seat when the valve is open, thereby permitting the seal element to flex radially. The rigidity in the closed position is

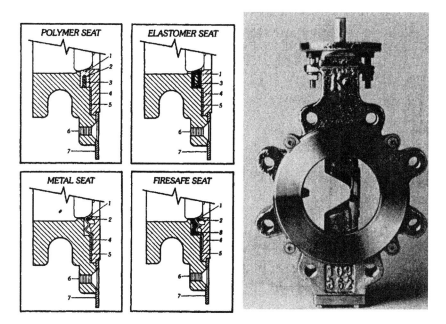

Figure 3-83. High-Performance Butterfly Valve Designed for Interference Seating. Scrap Views Show Plastic Seat, Elastomer Seat, Metal Seat, and Plastic/Metal Composition Seat for Flammable Liquid Service. (*Courtesy of Keystone International, Inc.*)

required to achieve the desired seat interference for a tight shut-off against high and low pressures, including vacuum.

Butterfly Valves for Fire Exposure

Butterfly valves, which may be exposed to plant fire when handling flammable fluids, must remain essentially fluid-tight internally and externally and be operable during and after a fire. These conditions may be met by fire-tested versions of high-performance butterfly valves.

The requirements for testing and evaluating the performance of butterfly valves when exposed to fire are similar to those for ball valves, described on page 108.

Body Configurations

The preferred body configuration for butterfly valves is the wafer, which is clamped between two pipeline flanges. An important advantage of this construction is that the bolts pulling the mating flanges together carry all the tensile stresses induced by the line strains and put the wafer in compression. This compressive stress is eased by the tensile stresses imposed by the internal fluid pressure. Flanged bodies, on the other hand, have to carry all the tensile stresses imposed by the line strains, and the tensile stresses from the line pressure are cumulative. This fact, together with the ability of most metals to handle compressive loads of up to twice their limit for tensile loads, strongly recommends the use of the wafer body.

However, if the downstream side of the butterfly valve serves also as a point of disconnection while the upstream side is still under pressure, the cross-bolted wafer body is unsuitable unless provided with a false flange. A flanged body or a lugged wafer body in which the lugs are threaded at each end to receive screws from the adjacent flanges is commonly used.

Torque Characteristic of Butterfly Valves

The torque required to operate butterfly valves consists of three main components:

T_h = hydrodynamic torque that is created by the flowing fluid acting on the disc

T_b = torque to overcome bearing friction

T_s = torque required to seat or unseat the disc

The hydrodynamic torque varies with the valve opening position and the pressure drop across the valve. In the case of symmetrical discs, this torque is identical for either direction of flow, and its direction of action is against the opening motion throughout. If the disc is offset, as in the disc shown in Figure 3-84, the hydrodynamic torque differs for each direction of flow, and the lowest torque develops when the flow is toward the disc. With flow toward the shaft, the torque acts against the opening motion throughout. However, with flow toward the disc, the torque acts only initially against the opening motion and then, with further valve opening, changes its directions.

The bearing, seating, and unseating torques, on the other hand, always act against the operating motion. The magnitude of the bearing torque corresponds to the resultant hydrodynamic force on the disc, while the magnitude of the seating and unseating torques is independent of flow.

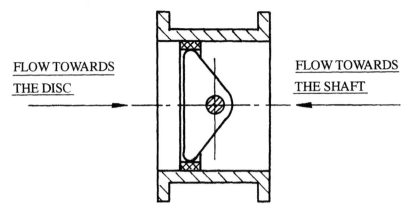

FLOW TOWARDS
THE DISC

FLOW TOWARDS
THE SHAFT

Figure 3-84. Offset Disc Configuration in Butterfly Valves.

In the case of interference-seated butterfly valves of sizes up to DN 400 (NPS 16), the seating and unseating torques normally dominate, provided the flow velocity is not too high. It is sufficient in this case to size the operating gear for these torques. However, the magnitude of the seating and unseating torques is influenced by the type of fluid handled and the operating frequency. For example, if the fluid has good lubricity and the valve is operated frequently, the seating and unseating torques are lower than for fluids that have little lubricity or consist of solids, or when the valve is operated infrequently. Manufacturers supply tables that give the seating and unseating torques for various conditions of operation.

In the case of interference-seated butterfly valves above DN 400 (NPS 16) and conditions of high-flow velocities, the hydrodynamic and bearing torques can greatly exceed the seating and unseating torques. The operator for such valves may therefore have to be selected in consultation with the manufacturer.

Figure 3-85 shows typical opening torque characteristics of butterfly valves with symmetrical and offset discs representing the summary of torques at a constant pressure loss across the valve. Under most operating conditions, however, the pressure drop across the valve decreases as the valve is being opened, as shown for example in Figure 3-86, for a given pumping installation. The decreasing pressure drop, in turn, decreases the corresponding operating torque as the valve opens. The maximum operating torque shifts thereby from the region of the fully open valve to the region of the partially open valve, as shown in Figure 3-87. This torque is

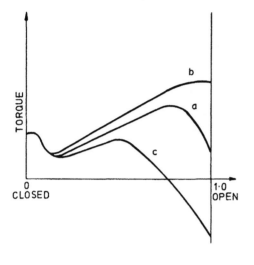

Figure 3-85. Opening Torque Characteristics of Butterfly Valves at Constant Pressure Drop Across Valve. (*Reprinted from Schiff and Hafen,*[40] *Courtesy of VAG-Armaturen GmbH.*) Curve a: Symmetrical Disc, Flow From Either Direction
Curve b: Off-Set Disc, Flow Towards Shaft
Curve c: Off-Set Disc, Flow Towards Disc

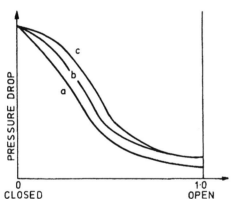

Figure 3-86. Pressure Drop Across Butterfly Valves for All Opening Positions in Actual Pumping Installation. (*Reprinted from Schiff and Hafen,*[40] *Courtesy of VAG-Armaturen GmbH.*) Curve a: Symmetrical Disc, Flow From Either Direction
Curve b: Off-Set Disc, Flow Towards Shaft
Curve c: Off-Set Disc, Flow Towards Disc

lowest if the disc is of the offset type and the direction of flow is toward the disc. However, this advantage is obtained at the penalty of a somewhat higher pressure drop. If the valve is large and the flow velocity high, this torque characteristic can be exploited to lower the cost of the operator.

Standards Pertaining to Butterfly Valves

Appendix C provides a list of U.S. and British standards pertaining to butterfly valves.

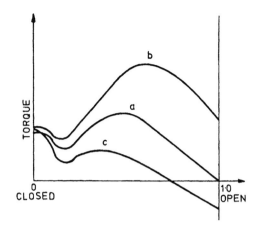

Figure 3-87. Opening Torque Characteristics of Butterfly Valves in the Actual Pumping Installation. (*Reprinted from Schiff and Hafen,*[40] *Courtesy of VAG-Armaturen GmbH.*)
Curve a: Symmetrical Disc, Flow From Either Direction
Curve b: Off-Set Disc, Flow Towards Shaft
Curve c: Off-Set Disc, Flow Towards D138isk

Applications

Duty:
  Stopping and starting flow
  Controlling flow
Service:
  Gases
  Liquids
  Slurries
  Powder
  Granules
  Sanitary handling of pharmaceuticals and food stuffs
  Vacuum

PINCH VALVES

Pinch valves are flex-body valves consisting of a flexible tube that is pinched either mechanically, as in the valves shown in Figure 3-88 through Figure 3-90, or by the application of a fluid pressure to the outside of the valve body, as in the valves shown in Figure 3-91 and Figure 3-92.

One of the principal advantages of this design concept is that the flow passage is straight without crevices and moving parts. The soft valve body also has the ability to seal around trapped solids. Pinch valves are therefore suitable for handling slurries and solids that would clog in obstructed flow passages, and for the sanitary handling of foodstuffs and pharmaceuticals. Depending on the construction material used for the valve body, pinch

Figure 3-88. Pinch Valve, Open Construction. (*Courtesy of Flexible Valve Corporation.*)

Figure 3-89. Pinch Valve, Enclosed Construction, and Valve Body 50% Prepinched for Sensitive Flow Control. Drain Plug May Be Removed for Vacuum Connection When Valve Is Used on Vacuum Service. (*Courtesy of RKL Controls, Inc.*)

Figure 3-90. Pinch Valve, Enclosed Construction, PTFE Valve Body Fitted with Tear Drops. (*Courtesy of Resistoflex Corporation.*)

valves will also handle severely abrasive fluids and most corrosive fluids.

Open and Enclosed Pinch Valves

Mechanically pinched valves may be of the open type, as in the valve shown in Figure 3-88, or of the enclosed type, as in the valves shown in Figure 3-89 and Figure 3-90.

The open construction is used if it is desirable to inspect the valve body visually and physically while the valve is in service. Bulges and soft spots that appear in the valve body are signs that the valve body needs to be replaced.

If the escape of the fluid from an accidentally ruptured valve body cannot be tolerated, the valve body must be encased. The casing is normally split along the axis of the flow passage for convenient access, but casings of unit construction are also being made.

Flow Control with Mechanically Pinched Valves

Pinch valves give little flow control between the fully open and the 50% pinched position because of the negligible pressure drop at these valve positions. Any further closing of the valve gives good flow control. Mechanically pinched valves for flow control duty are therefore often 50% prepinched.

If the fluid is severely erosive, flow control near the closed valve position must be avoided to prevent grooving of the valve body. For this reason also, pinch valves for erosive duty must always be closed fluid-tight to prevent grooving of the valve body as a result of leakage flow.

Flow Control with Fluid-Pressure Operated Pinch Valves

Fluid-pressure operated pinch valves are not normally suitable for manual flow control because any change in the downstream pressure will automatically reset the valve position. The flow control characteristic fluid-pressure operated pinch valves, such as the one shown in Figure 3-91, is otherwise similar to that of mechanically pinched valves.

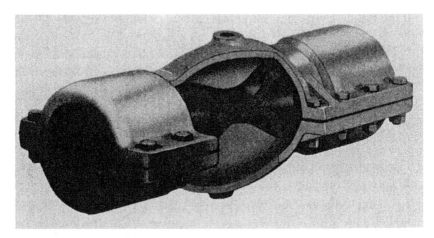

Figure 3-91. Pinch Valve, Fluid-Pressure Operated. (*Courtesy of RKL Controls, Inc.*)

THROTTLING ACTION

SIDE VIEW   END VIEW

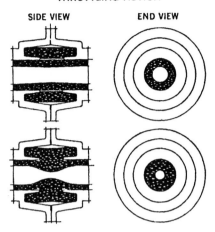

Figure 3-92. Pinch Valve, Fluid-Pressure Operated, with Iris-Like Closing Action. The Constricting Action of the Rubber Muscle upon the Sleeve Is Literally a 360° Squeeze. Pressure is Evenly Applied to the Circumference of the Sleeve—Hence the Always Round, Always Centered Hole. (*Courtesy of the Clarkson Company.*)

An exception to this flow control characteristic is the valve shown in Figure 3-92, in which the closing action of the valve body is iris-like. This closing action gives equal-percentage flow control throughout.

The iris-like closure allows the valve to pass larger particles through in any valve position than is possible with any other valve. This ability considerably reduces the tendency of some slurries to bridge in the partly closed valve. The circular throttling orifice tends to eliminate also the grooving of the valve body that develops in other types of pinch valves when severely throttling abrasive fluids. On the debit side, the valve cannot be fully closed.

Valve Body

The valve body is made of an elastomer, PTFE, or a similar material, or a combination of these materials. Fabric reinforcement may be used to increase the mechanical strength of the valve body.

Natural rubber is the best construction material for the valve body from the point of resistance to cracking from flexing. Natural rubber also shows excellent resistance to abrasive wear and to corrosion from many corrosive fluids.

If the fluid is chemically active, the surface of highly stressed rubber is more readily attacked than that of unstressed rubber. The corrosives may also form a thin, highly resistant film of corrosion products, which is much less flexible than the parent rubber. This film will crack on severe flexing and expose the rubber to further attack. Repeated flexing will finally lead to cracking right through the rubber.

Some of the synthetic rubbers are considerably less subject to this form of attack than natural rubber. These rubbers extend, therefore, the application of pinch valves for corrosive fluids. In developing such compounds, manufacturers aim not only at corrosion resistance but also at high tensile strength combined with softness for abrasion resistance and ease of closure.

To minimize the severity of flexing, the inside of the valve body may be provided with groove-like recesses on opposite sides, along which the valve body folds on closing. The valve shown in Figure 3-90 reduces the severity of flexing by providing tear drops inside the valve body on opposite sides. This design permits the valve body to be made entirely of PTFE.

Limitations

The flexible body puts some limitations on the use of pinch valves. These limitations may be overcome in some cases by special body designs or by the method of valve operation.

For example, fluid-pressure-operated pinch valve tend to collapse on suction duty. If mechanically pinched valves are used for this duty, the body must be positively attached to the operating mechanism.

Pinch valves on the downstream side of a pump should always be opened prior to starting up the pump. If the valve is closed, the air between the pump and the valve will compress and, upon cracking the valve, escape rapidly. The liquid column that follows will then hit the valve body with a heavy blow—perhaps severe enough to burst the valve body.

Pinch valves may also fail if the flow pulsates. The valve body will pant and finally fail due to fatigue.

There is a limitation when using pinch valves as the main shut-off valve in liquid-handling systems in which the liquid can become locked and has no room to move. Because the valve must be able to displace a certain amount of liquid when closing, the valve cannot be operated in this situation. Any effort to do so may cause the valve body to burst.

Standards Pertaining to Pinch Valves

National standards that apply specifically to pinch valves do not exist.

Applications

Duty:
  Stopping and starting flow
  Controlling flow
Service:
  Liquids
  Abrasive slurries
  Powders
  Granules
  Sanitary handling of pharmaceuticals and food stuffs

DIAPHRAGM VALVES

Diaphragm valves are flex-body valves in which the valve body consists of a rigid and flexible section. The flexible body section is provided by a diaphragm which, in connection with a compressor, represents the closure member. The seat is provided by the rigid body section and may consist of a weir across the flow passage, as in the valve shown in Figure 3-93, or be provided by the wall of a straight-through flow passage, as in Figure 3-94.

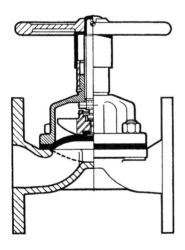
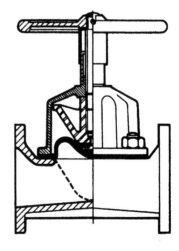

Figure 3-93. Diaphragm Valve, Weir Type. (*Courtesy of Saunders Valve Company Limited.*)

Figure 3-94. Diaphragm Valve, Straight-Through Type. (*Courtesy of Saunders Valve Company Limited.*)

Diaphragm valves share a similar advantage with pinch valves; namely, a flow passage that is not obstructed by moving parts and is free of crevices. They may, therefore, be put to uses similar to pinch valves, including the sanitary handling of foodstuffs and pharmaceuticals.

Weir-Type Diaphragm Valves

The weir in the flow passage is designed to reduce flexing of the diaphragm to a minimum, while still providing a smooth and streamlined flow passage. The flexing stress in the diaphragm is therefore minimal, resulting in a correspondingly long diaphragm life. The short stroke of these valves also permits the use of plastics such as PTFE for the diaphragm, which would be too inflexible for longer strokes. The back of such diaphragms is lined with an elastomer, which promotes a uniform seating stress upon valve closing.

Weir-type diaphragm valves may also be used in general and high vacuum service. However, some valve makers require a specially reinforced diaphragm in high vacuum service.

Because the diaphragm area is large compared with the flow passage, the fluid pressure imposes a correspondingly high force on the raised diaphragm. The resulting closure torque limits the size to which diaphragm valves can be made. Typically, weir-type diaphragm valves of the type shown in Figure 3-93 are made in sizes up to DN 350 (NPS 14). Larger weir-type diaphragm valves up to DN 500 (NPS 20) are provided with a double-bonnet assembly, as shown in Figure 3-95.

Weir-type diaphragm valves are also available with a body of T-configuration, as shown in Figure 3-96, in which a branch connects to the main flow passage without impeding pipeline flow. The main function of these valves is for sampling duty where the taking of a true sample from the flowing fluid must be assured.

The stem of the valve shown in Figure 3-93 is of the rising type. To protect the external stem thread from dust and immediate outside corrosive influences. The handwheel carries a shroud that covers the stem thread while sliding over a yellow lift-indicator sleeve. The yellow lift-indicator sleeve, in turn, carries a prepacked lubrication chamber to lubricate the stem thread for long life.

If required, the stem may be provided with an O-ring seal against the bonnet to prevent fluid from escaping into the surroundings of the valve should the diaphragm break in service.

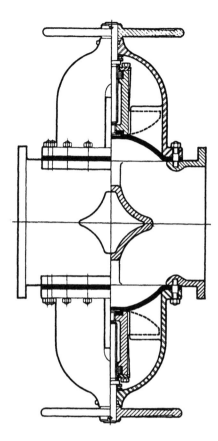

Figure 3-95. Diaphragm Valve, Weir Type, with Double-Bonnet Assembly in Connection with Large Valve Sizes. (*Courtesy of Saunders Valve Company Limited.*)

Conventional weir-type diaphragm valves may also be used in horizontal lines that must be self-draining. Self-draining is achieved by mounting the valve with the stem approximately 15° to 25° up from the horizontal, provided the horizontal line itself has some fall.

Straight-Through Diaphragm Valves

Diaphragm valves with a straight-through flow passage require a more flexible diaphragm than weir-type diaphragm valves. For this reason, the construction material for diaphragms of straight-through diaphragm valves is restricted to elastomers.

Because of the high flexibility and large area of these diaphragms, high vacuum will tend to balloon the diaphragm into the flow passage.

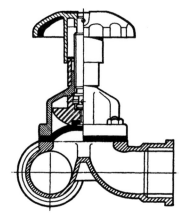

Figure 3-96. Diaphragm Valve, Weir Type, with T-Branch. (*Courtesy of Saunders Valve Company Limited.*)

The degree of ballooning varies thereby with make, causing either a small and acceptable reduction in flow area only or a collapse of the diaphragm. In the latter case, the bonnet must be evacuated to balance the pressures on the diaphragm. When using these valves for high vacuum, the manufacturer should be consulted.

Straight-through diaphragm valves are also available with full-bore and reduced-bore flow passage. In the case of reduced-bore valves, the bonnet assembly of the next smaller valve is used. For example, a DN 50 (NPS 2) reduced-bore valve is fitted with a DN 40 (NPS $1\frac{1}{2}$) bonnet. The construction of the bonnet is otherwise similar to that of weir-type diaphragm valves.

The size range of straight-through diaphragm valves typically covers valves up to DN 350 (NPS 14).

Construction Materials

The valve body of diaphragm valves is available in a great variety of construction materials, including plastics, to meet service requirements. The simple body shape also lends itself readily to lining with a great variety of corrosion-resistant materials, leading often to low-cost solutions for an otherwise expensive valve.

Because the diaphragm separates the bonnet from the flowing fluid, the bonnet is normally made of cast iron and epoxy coated inside and outside. If requested, the bonnet is available also in a variety of other materials.

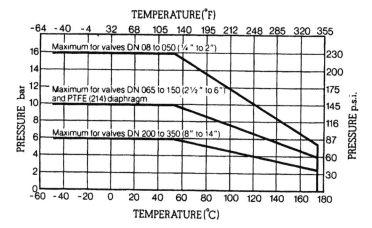

Figure 3-97. Pressure/Temperature Relationship of Weir-Type Diaphragm Valves. (*Courtesy of Saunders Valve Company Limited.*)

Diaphragms are available in a great variety of elastomers and plastics. Valve makers' catalogs advise the user on the selection of the diaphragm material for a given application.

Valve Pressure/Temperature Relationships

Figure 3-97 and Figure 3-98 show typical pressure/temperature relationships of weir-type and straight-through diaphragm valves. However, not all construction materials permit the full range of these relationships. The valve user must therefore consult the manufacturer's catalog for the permissible operating pressure at a given operating temperature.

Valve Flow Characteristics

Figure 3-99 shows the typical inherent and installed flow characteristics of weir-type diaphragm valves, though the shapes of the curves may vary to some extent between valve sizes. In the case of automatic control, these characteristics may be modified using a variable cam positioner.

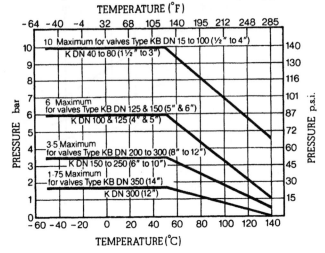

Figure 3-98. Pressure/Temperature Relationship of Straight-Through Diaphragm Valves. (*Courtesy of Saunders Valve Company Limited.*)

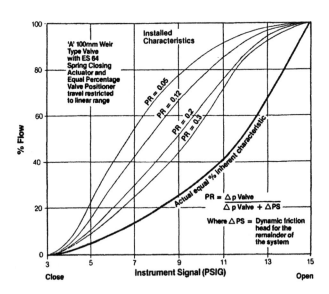

Figure 3-99. Inherent and Installed Flow Characteristics of Weir-Type Diaphragm Valves. (*Courtesy of Saunders Valve Company Limited.*)

Operational Limitations

Like pinch valves, diaphragm valves displace a certain amount of fluid when operated; that is, diaphragm valves are positive-displacement type valves. For this reason, diaphragm valves must not be installed as shut-off valves in lines containing an incompressible fluid where the fluid has no room to move.

Standards Pertaining to Diaphragm Valves

Appendix C provides a list of U.S., British and ISO standards pertaining to diaphragm valves.

Applications

Duty for weir-type and straight-through diaphragm valves:
    Stopping and starting flow
    Controlling flow
Service for weir-type diaphragm valves:
    Gases, may carry solids
    Liquids, may carry solids
    Viscous fluids
    Leak-proof handling of hazardous fluids
    Sanitary handling of pharmaceuticals and foodstuffs
    Vacuum
Service for straight-through diaphragm valves:
    Gases, may carry solids
    Liquid, may carry solids
    Viscous fluids
    Sludges
    Slurries, may carry abrasives
    Dry media
    Vacuum (consult manufacturer)

STAINLESS STEEL VALVES [41,42]

Corrosion-Resistant Alloys

The least corrosion-resistant alloy is normally thought of as steel AISI type 304 (18 Cr, 10 Ni). Stainless steel AISI type 316 (18 Cr, 12 Ni, 2.5 Mo) has a wider range of corrosion-resistance than type 304, and valve makers often endeavor to standardize on type 316 as the least corrosion-resistant alloy. If the valve is to be welded into the pipeline, the low carbon grades (less than 0.3% carbon) are better than stabilized grades. Flanged valves require welding only when a casting defect has to be repaired. Because the repair is done prior to the 1100°C (2000°F) water quench solution anneal, standard carbon grades are quite satisfactory for flanged valves.

Crevice Corrosion

Practically all corrosion-resistant alloys are susceptible to crevice corrosion. Good valve designs therefore avoid threading any component that comes in contact with the corrosive fluid. For this reason, valve seats are normally made an integral part of the valve body. An exception is body designs in which the seat is clamped between two body halves, as in the valve shown in Figure 3-12. However, the gaskets between the seat and the body halves must be cleanly cut to avoid crevices.

If the valve is to be screwed into the pipeline, seal welding will improve the performance of the screwed connection. Alternatively, thread sealants, which harden after application, are helpful in combating crevice corrosion in threaded joints. Flanged facings, which incorporate crevices such as tongue and groove, should be avoided.

The points of porosity in the valve body that are exposed to the corrosive fluid can likewise produce crevice corrosion. The body may thereby corrode through at the point of porosity and produce gross leakage, while the remainder of the body stays in good condition.

Galling of Valve Parts

Published information usually shows that stainless steel in sliding contact, particularly austenitic grades of like compositions, are susceptible to galling. This galling tendency diminishes considerably if the fluid has good lubricity and the seating surfaces can retain the lubricants and protective contaminants. Polished surfaces have only a limited ability to retain lubricants and contaminants and therefore display an increased tendency to gall when in sliding contact. For this reason, the seating faces of stainless steel valves are usually very finely machined rather than polished.

If a high galling tendency is expected, as when handling dry gases, one of the seating faces may be faced with stellite, which is known to provide good resistance to a wide range of corrosives.

Seizing of the valve stem in the yoke bush is commonly avoided by choosing dissimilar materials for the yoke bush and the stem. A material frequently used for the bush in stainless steel valves is Ni-Resist ductile iron D2, which provides complete freedom from galling due to the graphite in its structure. If the bush is made of stainless steel, the free-machining grade type 303 provides remarkable freedom from galling in conjunction with stainless steel grades type 304 and 316 for the stem.

Light-Weight Valve Constructions

Efforts in the United States to reduce the cost of stainless steel valves led to the development of standards for 150 lb light-weight stainless steel valves. The pressure ratings specified in these standards apply only to valves made from austenitic materials.

The flanges to these standards are thinner than the corresponding full-rating carbon steel flanges and have plain flat faces. Many users object to the light-weight flanges and request full-rating flanges with a raised face.

The light-weight bodies are, of course, more flexible than the bodies of full-rating carbon steel valves. This is particularly important for gate valves, in which body movements can unseat the disc. Experience has shown that plain solid wedges may be used only for sizes up to DN 100 (NPS 4). Larger valves are satisfactory only if the wedge is of the self-aligning type.

Standards Pertaining to Stainless Steel Valves

Appendix C provides a list of standards pertaining to stainless steel valves.

4

CHECK VALVES

FUNCTION OF CHECK VALVES

The prime function of a check valve is to protect mechanical equipment in a piping system by preventing reversal of flow by the fluid. This is particularly important in the case of pumps and compressors, where back flow could damage the internals of the equipment and cause an unnecessary shutdown of the system and in severe cases the complete plant.

Generally speaking check valves have no requirement for operators, and so the valve is operated automatically by flow reversal; however, in very special circumstances this uni-directional facility has to be overridden. Check valves either can be fitted with a device that allows the closure plate(s) to be locked open or alternatively can have the closure plate(s) removed. The latter alternative requires dismantling the valve, removing the plates, and re-installing the valve.

Check valves are automatic valves that open with forward flow and close against reverse flow.

This mode of flow regulation is required to prevent return flow, to maintain prime after the pump has stopped, to enable reciprocating pumps and compressors to function, and to prevent rotary pumps and compressors from driving standby units in reverse. Check valves may also be required in lines feeding a secondary system in which the pressure can rise above that of the primary system.

Grouping of Check Valves

Check valves may be grouped according to the way the closure member moves onto the seat. Four groups of check valves are then distinguished:

1. Lift check valves. The closure member travels in the direction normal to the plane of the seat, as in the valves shown in Figure 4-1 through Figure 4-7.
2. Swing check valves. The closure member swings about a hinge, which is mounted outside the seat, as in the valves shown in Figure 4-8 through Figure 4-10.
3. Tilting-disc check valves. The closure member tilts about a hinge, which is mounted near, but above, the center of the seat, as in the valve shown in Figure 4-11.
4. Diaphragm check valves. The closure member consists of a diaphragm, which deflects from or against the seat, as in the valves shown in Figure 4-12 through Figure 4-14.

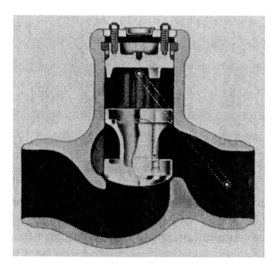

Figure 4-1. Lift Check Valve with Piston-Type Disc, Standard Pattern. (*Courtesy of Edward Valves Inc.*)

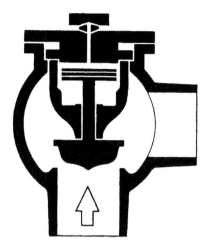

Figure 4-2. Lift Check Valve, Angle Pattern, with Built-in Dashpot, Which Comes Into Play During the Final Closing Movement. (*Courtesy of Sempell A.G.*)

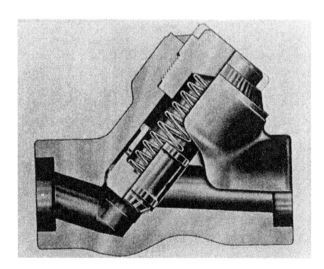

Figure 4-3. Lift Check Valve with Piston-Type Disc, Oblique Pattern. (*Courtesy of Edward Valves Inc.*)

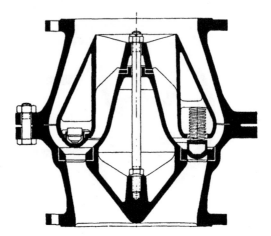

Figure 4-4. Lift Check Valve with Spring-Loaded Ring-Shaped Disc. (*Courtesy of Mannesmann-Meer AG.*)

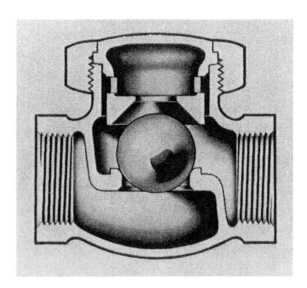

Figure 4-5. Lift Check Valve with Ball-Type Disc, Standard Pattern. (*Courtesy of Crane Co.*)

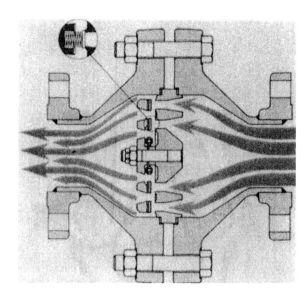

Figure 4-6. Lift Check Valve for Pulsating Gas Flow Characterized by Minimum Valve Lift, and Low Inertia and Frictionless Guiding of Closure Member. (*Courtesy of Hoerbiger Corporation of America.*)

Figure 4-7. Combined Lift Check and Stop Valve with Piston-Type Disc, Oblique Pattern. (*Courtesy of Edward Valves Inc.*)

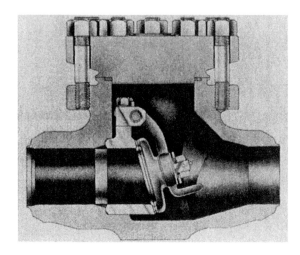

Figure 4-8. Swing Check Valve. (*Courtesy of Velan Engineering, Limited.*)

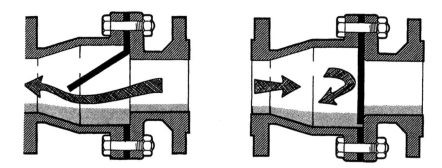

Figure 4-9. Swing Check Valve with Steel Reinforced Rubber Disc, the Disc Being an Integral Part of the Body Gasket. (*Courtesy of Saunders Valve Company Limited.*)

Operation of Check Valves

Check valves operate in a manner that avoids:

1. The formation of an excessively high surge pressure as a result of the valve closing.
2. Rapid fluctuating movements of the valve closure member.

Figure 4-10. Double-Disc Swing Check Valve. (*Courtesy of Mission Manufacturing Co.*)

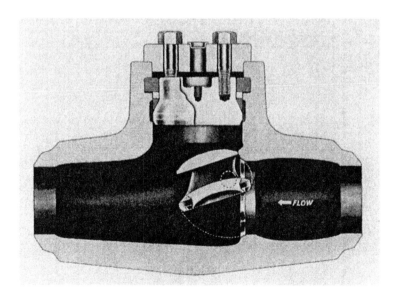

Figure 4-11. Tilting-Disc Check Valve. (*Courtesy of Edward Valves Inc.*)

To avoid the formation of an excessively high surge pressure as a result of the valve closing, the valve must close fast enough to prevent the development of a significant reverse flow velocity that on sudden shut-off is the source of the surge pressure.

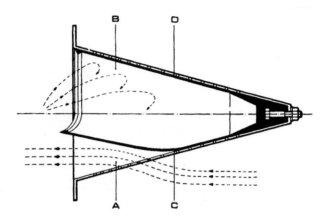

Figure 4-12. Diaphragm Check Valve with Cone-Shaped Diaphragm. (*Courtesy of Northvale Engineering.*)

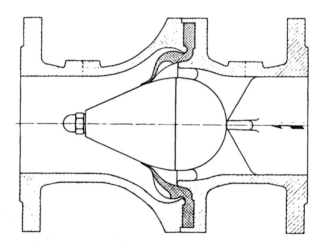

Figure 4-13. Diaphragm Check Valve with Ring-Shaped Pleated Diaphragm. (*Courtesy of VAG-Armaturen GmbH.*)

However, the speed with which forward flow retards can vary greatly between fluid systems. If, for example, the fluid system incorporates a number of pumps in parallel and one fails suddenly, the check valve at the outlet of the pump that failed must close almost instantaneously. On the other hand, if the fluid system contains only one pump that suddenly fails, and if the delivery line is long and the back pressure at the outlet of the

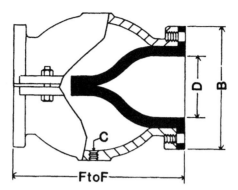

Figure 4-14. Diaphragm Check Valve, Incorporating Flattened Rubber Sleeve Closure Member. (*Courtesy of Red Valve Company Inc.*)

pipe and the pumping elevation are low, a check valve with a slow closing characteristic is satisfactory.

Rapid fluctuating movements of the closure member must be avoided to prevent excessive wear of the moving valve parts, which could result in early failure of the valve. Such movements can be avoided by sizing the valve for a flow velocity that forces the closure member firmly against a stop. If flow pulsates, check valves should be mounted as far away as practical from the source of flow pulsations. Rapid fluctuations of the closure member may also be caused by violent flow disturbances. When this situation exists, the valve should be located at a point where flow disturbances are at a minimum.

The first step in the selection of check valves, therefore, is to recognize the conditions under which the valve operates.

Assessment of Check Valves for Fast Closing[43]

In most practical applications, check valves can be assessed only qualitatively for fast closing speed. The following criteria may serve as a guide:

1. Travel of the closure member from the fully open to the closed position should be as short as possible. Thus, from the point of speed of closing, a smaller valve is potentially faster closing than a larger valve of otherwise the same design.

2. The inertia of the closure member should be as low as possible, but the closing force should be appropriately high to ensure maximum response to declining forward flow. From the point of low inertia, the closure member should be of light construction. To combine light-weight construction with a high closing force, the closing force from the weight of the closure member may have to be augmented by a spring force.
3. Restrictions around the moving closure member that retard the free closing movement of the closure member should be avoided.

Application of Mathematics to the Operation of Check Valves

The application of mathematics to the operation of check valves is of relatively recent origin. Pool, Porwit, and Carlton[40] describe a calculation method for check valves with a hinged disc that involves setting up the equation of motion for the disc and applying to that the deceleration characteristic of the flowing fluid within the system. Before the equation of motion for the disc can be written, certain physical constants of the valve must be known. The calculation determines the reverse flow velocity at the instant of sudden shut-off. The surge pressure due to the sudden shut-off of the reverse flow can then be calculated as described in Chapter 2.

It is important for the valve user to know that valve manufacturers can use mathematics in designing check valves for given critical applications and predicting surge pressure.

DESIGN OF CHECK VALVES

Lift Check Valves

The check valves shown in Figure 4-1 through Figure 4-7 represent a cross section of the family of lift check valves.

Lift check valves have an advantage over most other types of check valves in that they need only a relatively short lift to obtain full valve opening. This lift is a minimum in lift check valves in which the flow passage at the seat is ring-shaped, as in the valves shown in Figure 4-4 and Figure 4-6. Lift check valves are, therefore, potentially fast closing.

In the majority of lift check valves, the closure member moves in a guide to ensure that the seatings mate on closing. However, such guiding also has a disadvantage in that dirt entering the guide can hang up the closure member, and viscous liquids will cause lazy valve operation or even cause the closure member to hang up. These types of lift check valves are therefore suitable for low viscosity fluids only, which are essentially free of solids. Some designs overcome this disadvantage, as in the valve shown in Figure 4-5, in which the closure member is ball-shaped and allowed to travel without being closely guided. When the valve closes, the ball-shaped closure member rolls into the seat to achieve the required alignment of the seatings.

The check valve shown in Figure 4-2 is specifically designed for applications in which a low surge pressure is critical. This is achieved in two ways, first, by providing the closure member with a conical extension that progressively throttles the flow as the valve closes, and second, by combining the closure member with a dashpot that comes into play in the last closing moments. A spring to assist closing of the valve has been purposely omitted, as breakage of the spring was considered a hazard for the service for which the valve is intended.

The check valve shown in Figure 4-6 is designed for gas service only. Depending on flow conditions, the valve may serve either as a constant-flow check valve in which the valve remains fully open in service irrespective of minor flow fluctuations, or as a pulsating-flow check valve in which the valve opens and closes with each pulse of the flowing gas.

Constant-flow check valves are used after centrifugal, lobe-type, and screw compressors, or after reciprocating compressors if the flow pulsations are low enough not to cause plate flutter. Pulsating-flow check valves are used after reciprocating compressors if the flow pulsations cause the valve to open and close with each pulsation. The valves are designed on the same principles as compressor valves and, therefore, are capable of withstanding the repeated impacts between the seatings. The manufacturer will advise whether a constant-flow or pulsating-flow check valve may be used for a given application.

The valves owe their operational characteristics to their design principle, based on minimum valve lift in conjunction with multiple ring-shaped seat orifices, low inertia of the plate-like closure member, frictionless guiding of the closure member, and the selection of a spring that is appropriate for the operating conditions.

The valve shown in Figure 4-7 is a combined lift check and stop valve. The valve resembles an oblique pattern globe valve in which the closure member is disconnected from the stem. When the stem is raised, the valve acts as a lift check valve. When the valve is lowered and firmly pressed against the closure member, the valve acts as a stop valve.

Lift check valves must be mounted in a position in which the weight of the closure member acts in the closing direction. Exceptions are some spring-loaded low-lift check valves in which the spring force is the predominant closing force. For this reason, the valves shown in Figure 4-1 and Figure 4-5 may be mounted in the horizontal flow position only, while the valve shown in Figure 4-2 may be mounted in the vertical upflow position only. The valves shown in Figure 4-3, Figure 4-4, and Figure 4-7 may be mounted in the horizontal and vertical upflow positions, while the valve shown in Figure 4-6 may be mounted in any flow position, including vertical downflow.

Swing Check Valves

Conventional swing check valves are provided with a disc-like closure member that swings about a hinge outside the seat, as in the valves shown in Figure 4-8 and Figure 4-9. Travel of the disc from the fully open to the closed position is greater than in most lift check valves. On the other hand, dirt and viscous fluids cannot easily hinder the rotation of the disc around the hinge. In the valve shown in Figure 4-9, the closure member is an integral part of the rubber gasket between the valve body halves. It is steel-reinforced, and opens and closes by bending a rubber strip connecting the closure member and the gasket.

As the size of swing check valves increases, weight and travel of the disc eventually become excessive for satisfactory valve operation. For this reason, swing check valves larger than about DN600 (NPS 24) are frequently designed as multi-disc swing check valves, and have a number of conventional swing discs mounted on a multi-seat diaphragm across the flow passage in the valve.

Swing check valves should be mounted in the horizontal position, but may also be mounted in the vertical position, provided the disc is prevented from reaching the stalling position. In the latter case, however, the closing moment of the disc, due to its weight, is very small in the fully open position, so the valve will tend to close late. To overcome slow response to retarding flow, the disc may be provided with a lever-mounted weight or spring loaded.

The check valve shown in Figure 4-10 is a double-disc swing check valve with two spring-loaded D-shaped discs mounted on a rib across the valve bore. This design reduces the length of the path along which the center of gravity of the disc travels; it also reduces the weight of such a disc by about 50%, compared with single-disc swing check valves of the same size. Coupled with spring loading, the response of the valve to retarding flow is therefore very fast.

Tilting-Disc Check Valves

Tilting-disc check valves such as the one shown in Figure 4-11 have a disc-like closure member that rotates about a pivot point between the center and edge of the disc and is offset from the plane of the seat. The disc drops thereby into the seat on closing, and lifts out of the seat on opening. Because the center of gravity of the disc halves describes only a short path between the fully open and the closed positions, tilting-disc check valves are potentially fast closing. This particular valve is, in addition, spring-loaded to ensure quick response to retarding forward flow.

Reference may be made also to the valve shown in Figure 3-77 that can serve as a butterfly valve, a tilting-disc check valve, or a combined tilting-disc check and stop valve, depending on the design of the drive.

Tilting-disc check valves have the disadvantage of being more expensive and also more difficult to repair than swing check valves. The use of tilting-disc check valves is therefore normally restricted to applications that cannot be met by swing check valves.

Diaphragm Check Valves

Diaphragm check valves such as those shown in Figure 4-12 through Figure 4-15 are not as well-known as other check valves, but they deserve attention.

The check valve shown in Figure 4-12 consists of a perforated cone-shaped basket that supports a matching diaphragm. This assembly is mounted in the pipeline between two flanges or clamped between pipe unions. Flow passing through the cone lifts the diaphragm off its seat and lets the fluid pass. When forward flow ceases, the diaphragm regains its original shape and closure is fast. One application worth mentioning is

Figure 4-15. Diaphragm Check Valve Incorporating Flattened Rubber Sleeve Closure Member, Used as Tide Gate. (*Courtesy of Red Valve Company, Inc.*)

in purge-gas lines, which feed into lines, handling slurry or gluey substances. Under these conditions, diaphragm valves tend to operate with great reliability, while other valves hang up very quickly.

The check valve shown in Figure 4-13 uses a closure member in the form of a pleated annular rubber diaphragm. When the valve is closed, a lip of the diaphragm rests with the pleats closed against a core in the flow passage. Forward flow opens the pleats, and the lip retracts from the seat. Because the diaphragm is elastically strained in the open position, and travel of the lip from the fully open to the closed position is short, the diaphragm check valve closes extremely fast. This valve is well-suited for applications in which the flow varies within wide limits. However, the pressure differential for which the valve may be used is limited to 10 bar (145 lb/in^2), and the operating temperature is limited to about 74°C (158°F).

The closure member of the diaphragm check valve shown in Figure 4-14 consists of a flexible sleeve that is flattened at one end. The flattened end of the sleeve opens on forward flow and recloses against reverse flow.

The sleeve is made in a large variety of elastomers, and is externally reinforced with plies of nylon fabric similar in construction to an automobile tire. The inside of the sleeve is soft and capable of embedding trapped solids. The valve is therefore particularly suitable for services in which the fluid carries solids in suspension or consists of a slurry.

Figure 4-15 shows an interesting application of this check valve as a tidal gate.

The valve is available in sizes as small as DN 3 (NPS 1/8) and as large as DN 3000 (NPS 120) for tidal gates.

Dashpots

The purpose of dashpots is to dampen the movement of the closure member.

The most important application of dashpots is in systems in which flow reverses very fast. If the check valve is unable to close fast enough to prevent a substantial reverse-flow buildup before sudden closure, a dashpot, designed to come into play during the last closing movements, can considerably reduce the formation of surge pressure.

SELECTION OF CHECK VALVES

Most check valves are selected qualitatively by comparing the required closing speed with the closing characteristic of the valve. This selection method leads to good results in the majority of applications. However, sizing is also a critical component of valve selection, as discussed in the following. If the application is critical, a reputable manufacturer should be consulted.

Check Valves for incompressible Fluids

These are selected primarily for their ability to close without introducing an unacceptably high surge pressure due to the sudden shut-off of reverse flow. Selecting these for a low pressure drop across the valve is normally only a secondary consideration.

The first step is qualitative assessment of the required closing speed for the check valve. Examples of how to assess the required closing speed in pumping installations are given in Chapter 2, page 41.

The second step is the selection of the type of check valve likely to meet the required closing speed, as deduced from page 151.

Check Valves for Compressible Fluids

Check valves for compressible fluids may be selected on a basis similar to that described for incompressible fluids. However, valve flutter can be a problem for high lift check valves in gas service, and the addition of a dashpot may be required.

Where rapidly fluctuating gas flow is encountered, compressor-type check valves such as that shown in Figure 4-6 are a good choice.

Standards Pertaining to Check Valves

Appendix C provides a list of USA and British standards pertaining to check valves.

5

PRESSURE RELIEF VALVES

PRINCIPAL TYPES OF PRESSURE RELIEF VALVES

Pressure relief valves are designed to protect a pressure system against excessive normal or subnormal pressure in the event of positive or negative excursion of the system pressure. They are required to open at a predetermined system pressure, to discharge or let enter a specified amount of fluid so as to prevent the system pressure from exceeding a specified normal or subnormal pressure limit, and to reclose after the normal system pressure has been restored. Pressure relief valves must also be self-actuated for maximum reliability except where permitted by the applicable Code of Practice for specific applications.

The most commonly employed pressure relief valve type is the direct-loaded pressure relief valve in which the fluid pressure acting on the valve disc in the opening direction is opposed by a direct mechanical-loading device such as a weight or spring. Direct-loaded pressure relief valves may also be provided with an auxiliary actuator, which may be designed to introduce a supplementary seating load on valve closing and/or to assist valve opening. The power for actuating the auxiliary actuator is commonly of an extraneous source. Should the external power source fail, the valves will open and close without hindrance in the manner of direct-loaded pressure relief valves.

The second type of pressure relief valve is the pilot-operated pressure relief valve that consists of a main valve and a pilot. The main valve is the actual pressure-relieving device while the pilot positions the disc of the main valve disc in response to the system pressure.

Traditionally in the U.S., direct-loaded pressure relief valves mainly intended for boiler and steam applications are referred to as *safety valves*, those mainly for liquid applications as *relief valves*, and those intended for both compressible and incompressible fluids as *safety relief valves*. This terminology is also applied in this book. None of these terms, however, fully describes the design or function of direct-loaded pressure relief valves, and the traditional terminology has been replaced in ASME Code and API RP 520 by the more generic term *pressure relief valve*.

Pilot-operated pressure relief valves in which the pilot is a self-actuated device are referred to in the ASME Code as pilot-operated pressure relief valves and those in which the pilot is an externally powered device as power-operated pressure relief valves.

The terminology used in the emerging European standards on pressure relief valves[1] differs to some extent from traditional U.S. terminology. In these standards, the term *safety valve* is a generic term for self-actuated reclosing pressure relief devices and covers direct-loaded and pilot-operated safety valves. Power-operated pressure relief valves are referred to as controlled safety pressure relief systems.

Pressure relief valves are covered in this book under the following headings:

- Direct-loaded pressure relief valves with subheadings safety valves, safety relief valves, liquid relief valves, vacuum relief valves, and direct-loaded pressure relief valves with auxiliary actuator.
- Piloted-pressure relief valves, consisting of a main valve in combination with a pilot that positions the disc of the main valve in response to the system pressure.

The construction, application, and sizing of pressure relief valves is subject to constraints by codes, or to agreement by statutory authorities that must be followed by the valve user.

---

[1] The emerging European standards for safety devices are labeled EN 1268, Parts 1 to 7. At the time of writing this book, the standards were still in draft form.

An overview of the ASME Code safety valve rules by M. D. Bernstein[1] and R. G. Friend[2] may be found in Reference 44. A reprint of a summary of the ASME Safety Valve Rules taken from this reference publication may be found in Appendix A of this book.

TERMINOLOGY

Pressure Relief Valves

Pressure relief device. A device designed to open in response to excessive internal normal or subnormal fluid pressure, and sized to prevent the fluid pressure from exceeding a specified normal or subnormal limit. The device may be either of the type that closes after the pressure excursion has receded or of the non-reclosing type.

Pressure relief valve. A pressure relief device that recloses automatically after the pressure excursion has receded.

Direct-loaded pressure relief valve. A pressure relief valve in which the fluid pressure acting on the disc in the opening direction is opposed by a direct mechanical-loading device such as a weight or a spring.

Safety valve. A direct-loaded pressure relief valve that is intended mainly for gas, vapor, or boiler and steam applications and characterized by pop-opening action.

Relief valve. A direct-loaded pressure relief valve intended mainly for liquid service.

Safety relief valve. A direct-loaded pressure relief valve that may be used either in gas or vapor service or in liquid service. The valve will open in gas or vapor service with a pop action, and in liquid service in proportion to the rise in overpressure though not necessarily linearly.

Conventional safety relief valve. A pressure relief valve in which the loading device is enclosed in a pressure-tight bonnet that is vented to the valve outlet.

Balanced safety relief valve. A pressure relief valve that incorporates a means for minimizing the effect of back pressure on the performance characteristic.

---

[1] Foster Wheeler Energy Corporation, Perryville Corporate Park, Clinton, NJ 08809.
[2] Crosby Valve & Gage Company, Wrentham, MA 02093.

Supplementary loaded pressure relief valve. A direct-loaded pressure relief valve with an externally powered auxiliary loading device that is designed to inhibit valve simmer by introducing a supplementary seating load during normal operation of the pressure system. Depending on the Code of Practice, the supplementary-loading force augmenting the spring force may be restricted or unrestricted. The load is automatically removed at a pressure not greater than the set pressure of the valve.[1]

Assisted pressure relief valve. A direct-loaded pressure relief valve with an externally powered auxiliary assist device that is designed to inhibit valve chatter by introducing a supplementary lifting force when the valve is called upon to open. When the safe operating pressure is restored, the assist device is deactivated and the valve closes normally. If the power to the assist device should fail, there is no interference with the normal operation of the valve.[2]

Pilot-operated pressure relief valve. A pressure relief valve consisting of a main valve that is the actual pressure relieving device, and a self-actuated pilot that controls the opening and closing of the main valve by either pressurizing or depressurizing the dome of the main valve.[3]

Pilot-operated pressure relief valve with direct-acting pilot. A pilot-operated pressure relief valve in which the pilot represents a pressure-actuated three-ported valve that controls the operation of the main valve.[4]

Pilot-operated pressure relief valve with indirect-acting pilot. A pilot-operated pressure relief valve in which the pilot represents a spring-loaded pressure relief valve that controls the operation of the main valve by the fluid being discharged.[5]

Power-actuated pressure relief valve. A pressure relief valve consisting of a main valve that is the actual pressure-relieving device and an externally powered piloting device that controls the opening and closing of the main valve by either pressurizing or depressurizing the dome of the main valve.[6]

---

[1] U.S. Code of Practice: ASME, Sect. III, NC 7511.3. European Standards: EN 1268-1 and 5.

[2] U.S. Code of Practice: ASME, Sect. III, NC-7512.1. European Standard: EN 1268-1.

[3] European Standard: EN 12688-1.

[4] European Standard: EN 12688-4.

[5] European Standard: EN 1268-1.

[6] U.S. Code of Practice: ASME, Sections I and III. European Standard: EN 1268-5.

Dome. The chamber at the top of the piston-shaped closure member of the main valve that is pressurized or depressurized to open or close the main valve.

Dimensional Characteristics

Discharge area. The controlled minimum net area that determines the flow through the valve.

Effective discharge area. The nominal or computed discharge area of a pressure relief valve used in recognized flow formulae in conjunction with the correlated effective coefficient of discharge to determine the size of the valve. The effective discharge area is less than the actual discharge area.

Huddling chamber. The annular chamber located beyond the seat of direct-loaded pressure relief valves formed by the lip around the disc for the purpose of generating pop-opening action in gas or vapor service, and raising the lifting force in both gas and liquid relief valves.

Secondary orifice. The annular opening at the outlet of the huddling chamber.

System Characteristics

Operating pressure. The maximum pressure that is expected during normal operating conditions.

Operating pressure differential. The pressure differential between operating pressure and set pressure expressed as a percentage of the set pressure or in pressure units.

Maximum allowable working pressure. The maximum pressure at which the pressure system is permitted to operate under service conditions in compliance with the applicable construction code for the pressure system. This pressure is also the maximum pressure setting of the pressure-relief devices that protect the pressure system.

Accumulation. The pressure increase in the pressure system over the maximum allowable working pressure, expressed in pressure units or percent of the maximum allowable working pressure.

Maximum allowable accumulation. The maximum allowable pressure increase in the fluid pressure over the maximum allowable working pressure as established by applicable codes for operating and fire contingencies, expressed in pressure units or percent of the maximum allowable working pressure.

Overpressure. The pressure increase over the set pressure of the relieving device, expressed as a percentage of the set pressure or in pressure units.

Device Characteristics

Actual coefficient of discharge K_d.[1] The ratio of the measured relieving capacity to the theoretical relieving capacity of a theoretical ideal nozzle.

Rated or certified coefficient of discharge K.[2] The actual coefficient of discharge Berated by a factor of 0.9.

Actual discharge capacity. The calculated discharge capacity of a theoretically perfect nozzle having a cross-sectional flow area equal to the discharge area of the valve multiplied by the actual coefficient of discharge.

Rated or certified discharge capacity. The discharge capacity of a theoretically perfect nozzle having a cross-sectional flow area equal to the flow area of the valve multiplied by the rated or certified coefficient of discharge.

Set pressure. The inlet pressure at which the pressure relief valve commences to open under service conditions.[3]

Start-to-open pressure. See "set pressure."

Popping pressure. The pressure at which a safety or safety relief valve pops open on gas or vapor service.

Opening pressure difference. The difference between set pressure and the popping pressure.

---

[1] ASME Code (1998) Section VIII, Division 1, UG-131 (e)(2).

[2] ASME Code (1998) Section VIII, Division 1, UG-131 (e)(3).

[3] The *Pressure Relief Device Certifications* book of The National Board of Boiler and Pressure Vessel Inspectors employs a variety of definitions of the term "set pressure," such as case 1: First leakage/initial discharge/simmer, case 2: First steady stream/first measurable lift, case 3: pop. The definition of the term *set pressure* as used in this book corresponds with the term "First measurable lift."

Cold differential test pressure. The pressure at which the pressure relief valve is adjusted on the test stand. The cold differential test pressure includes corrections for service conditions of back pressure or temperature or both.

Relieving pressure. The set pressure plus overpressure.

Overpressure. The pressure increase over the set pressure, expressed as a percentage of the set pressure or in pressure units.

Back pressure. The pressure that exists at the outlet of a pressure relief valve as a result of the pressure in the discharge system. It is the sum of built-up and superimposed back pressures, expressed as a percentage of the set pressure or in pressure units.

Built-up back pressure. The increase in pressure in the discharge header that develops as a result of flow after the pressure relief valve opens, expressed as a percentage of the set pressure or in pressure units.

Superimposed back pressure. The pressure that is present at the valve outlet when the valve is required to open, expressed as a percentage of the set pressure or in pressure units. The superimposed back pressure may be constant or variable with time.

Secondary back pressure. The back pressure component due to flow within the valve body to the valve outlet, normally a function of the valve orifice area compared to the valve outlet area and the shape of the body interior downstream of the nozzle.

Reseating pressure. The pressure at receding overpressure at which the valve disc reestablishes contact with the nozzle.

Blowdown. The difference between the set pressure and the pressure at resealing of the pressure relief valve, expressed as a percentage of the set pressure.[1]

Valve blowdown setting. The difference between the set pressure and the valve inlet pressure at the commencement of rapid valve closing, being the sum of the blowdown of the seated valve and the valve inlet pressure loss at the commencement of rapid valve closing, expressed as a percentage of the set pressure.

---

[1] In pop-opening pressure relief valves, blowdown is the difference between the valve blowdown setting and the inlet pressure loss at the commencement of rapid reclosing. In modulating pressure relief valves, blowdown is due to dynamic friction within the body of direct acting valves or within the pilot of pilot operated valves with direct acting pilot.

Blowdown pressure. See "reseating pressure."

Flutter. Abnormal, rapid reciprocating motion of the movable parts of the pressure relief valve in which the disc does not contact the nozzle.

Chatter. Abnormal, reciprocating motion of the movable parts of the pressure relief valve in which the disc contacts the nozzle.

Crawl. The gradual decreasing of the set pressure of spring-loaded pressure relief valves from below to normal after the temperature of the spring has been raised by the fluid just discharged.

Simmer. The audible or visible escape of compressible fluid between the seat and the disc at no measurable capacity as the set pressure is approached.

DIRECT-LOADED PRESSURE RELIEF VALVES

Review

Early pressure relief valves were of the direct-loaded type in which the disc was loaded by a weight. The shape of the disc resembled that of a globe valve and was commonly either tapered or flat. When rising fluid pressure raised the disc off its seat, the escaping fluid was not able to impart much of its kinetic energy onto the disc. Consequently, the lift of the disc was very small within the allowable overpressure. Early efforts were therefore directed to improve the valve lift.

An Englishman, Charles Ritchie, achieved the first significant improvement in valve lift in 1848 with a valve that exploited the expansive property of the gas for raising the disc. This was achieved by providing the disc with a peripheral flow deflector, which, together with a lip around the seal, formed an annular chamber with a secondary orifice around the seat, as shown in Figure 5-1. When a valve thus designed begins to open, the discharging gas expands in the annular chamber, but cannot readily escape. Therefore, the static pressure in this chamber rises sharply and, acting now also on an enlarged area on the underside of the disc, causes the valve to open suddenly. But as the escaping gas deflects on the disc only through around 90°, only a portion of its kinetic energy is converted into lifting force so that the valve can open only partially within the normally permissible overpressure. When the overpressure recedes, the static pressure in the annular chamber builds up again, causing the disc initially to huddle above the seat until the operating pressure has dropped to below the set

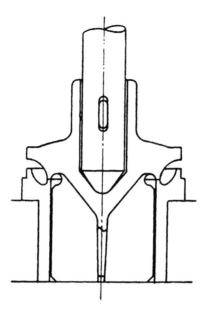

Figure 5-1. Valve Elements of Pressure Relief Valve, Ritchie Type.

pressure. The resulting difference between the set pressure and reseating pressure is referred to as the blowdown. Liquids, being incompressible, cannot develop a sudden pressure rise in the annular chamber, and the valve initially will open only a little with rising overpressure.

William Naylor introduced in 1863 an improved lift pressure relief valve in which a lip around the disc turned the discharging fluid through 180°, as shown in Figure 5-2. By this construction, the flowing fluid was able to impart the maximum lifting force on the disc from its momentum. But as the valve can open only in proportion to the rising overpressure, flow rate and the corresponding lifting force were initially too small to raise the disc significantly within the permissible overpressure.

Modern designs combine the principles of the Ritchie and Taylor valves; namely, they include the provision of a lip around the disc that is designed to form an annular chamber with a secondary orifice around the seat and to deflect the discharging fluid through about 180°. Over time, liquid relief valves have also been developed based on this design principle that open fully within an overpressure of 10%.

All early pressure relief valves employed a weight for loading the disc for two reasons. First, it was difficult at that time to produce a satisfactory

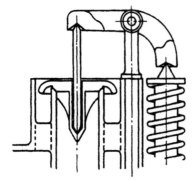

Figure 5-2. Pressure Relief Valve, Naylor Type.

spring. Second, objections were raised against the spring characteristic to raise the loading as the disc rises.

However, a weight soon becomes very heavy and eventually impractical as valve size and operating pressure increase. Lever mounting of the weight extends the range of application, but the lever cannot conveniently be enclosed to prevent unauthorized interference with the pressure setting.

Spring loading overcomes these disadvantages. The spring characteristic can also be matched to the lifting force characteristic that arises from the fluid acting on the disc. For these reasons, nearly all direct-loaded pressure relief valves are now spring loaded. Exceptions are pressure relief valves for low-pressure duties, depending on Code of Practice.

The following describes a typical range of direct-loaded pressure relief valves as offered by the industry.

Safety Valves

Safety valves, such as those shown in Figure 5-3 through Figure 5-6, are intended primarily for the relief of steam in industrial boiler plants and other steam systems. Their design is directed towards protecting the spring from excessive temperature rises that might cause drift of the spring setting and possibly spring relaxation over time. To provide this protection, safety valves are commonly provided with an open bonnet that allows steam leaking into the bonnet to escape directly into the atmosphere around the valve.

The valve shown in Figure 5-4 offers a further protection to spring overheating by mounting a lantern ring between valve body and bonnet.

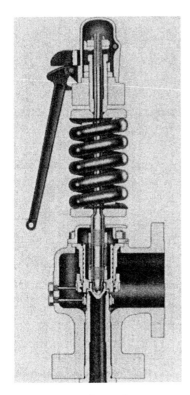 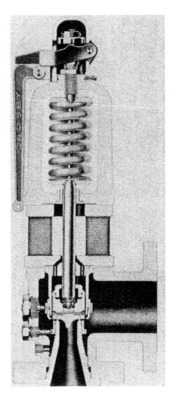

Figure 5-3. Safety Valve.
(*Courtesy of Dresser Industries.*)

Figure 5-4. Safety Valve
With Cooling Spool. (*Courtesy
of Crosby Valve & Gage
Company, Wrentham, MA.*)

The valve shown in Figure 5-6 is unique in that it incorporates an eductor formed by channels around the disc. When the valve pops open, the eductor becomes active and evacuates the chamber above the disc. The purpose of the eductor is to assist sharp opening and closing of the valve.

The disc of safety valves is commonly guided in a sleeve that is provided with a screwed ring for changing the direction of deflection of the escaping fluid and, in turn, for changing the reactive force acting on the underside of the disc. This method of adjusting the reactive force allows sensitive blowdown adjustment. Lowering the guide ring lengthens the blowdown. Raising the guide ring shortens the blowdown.

The second ring below the nozzle seat is referred to as the nozzle ring. Its main purpose is to control the difference between set pressure and popping pressure. Raising the nozzle ring reduces the width of the secondary orifice

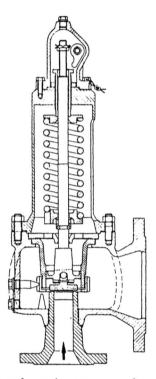

Figure 5-5. Safety Valve. (*Courtesy of Sempell A.G.*)

that leads to an earlier popping of the valve but, in turn, lengthens the blowdown by some small amount. Conversely, lowering the nozzle ring increases the difference between set pressure and popping pressure but, in turn, lengthens blowdown by some small amount.

The lower blowdown ring is commonly factory set and should not normally need readjustment when putting a new valve into service. No attempt should be made to eliminate the difference between set pressure and popping pressure altogether.

The valve shown in Figure 5-5 has done away with the nozzle ring. The width of the secondary orifice is fixed in this case by the geometry of seat and disc.

Safety valves with open bonnets are partially balanced by the stem guide. This may allow safety valves to operate against a built-up back pressure of about 20%, but the manufacturer must be consulted. They are, however, not suitable for superimposed back pressure because of leakage

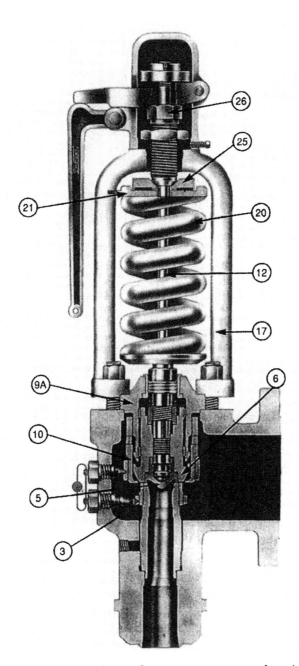

Figure 5-6. Safety Valve with Eductor Lift Assistance. (*Courtesy of Crosby Valve & Gage Company, Wrentham, MA.*)

around the stem to the bonnet. For this reason, safety valves should not be manifolded.

Safety Relief Valves

Safety relief valves are general-purpose pressure relief valves for use in either gas or liquid service as encountered in the process industry. Because the fluids handled in these industries cannot normally be tolerated to escape around the valve, the bonnet of these valves is either vented to the valve outlet or sealed against the valve chamber by means of bellows. An exception is safety relief valves for process steam duty in which the bonnet is provided with windows as in safety valves.

When used on gas or vapor service, the valves pop open. When used on liquid service, the valves modulate open and reach the fully open position at an overpressure of about 25% or, in newer designs, at an overpressure of 10%, as in the valves shown in Figure 5-7 and Figure 5-10.

The industry distinguishes between conventional safety relief valves and balanced safety relief valves.

Conventional safety relief valves. Conventional safety relief valves are provided with a closed bonnet that is vented to the valve outlet, as in the valves shown in Figure 5-7 and Figure 5-8. By this design, back pressure is allowed to act on the entire back of the disc.

Prior to valve opening, the forces from back pressure acting on both sides of that portion of the disc that overhangs the nozzle balance each other while the area of the back covering the valve nozzle is unbalanced. Superimposed back pressure acting on the unbalanced area thus raises the set pressure by the amount of the superimposed back pressure. The bench pressure setting must therefore be lowered by the amount of superimposed back pressure to allow the valve to open at the operational set pressure. For this reason, conventional safety relief valves are not suitable for variable superimposed back pressure and should not normally be manifolded.

Upon valve opening, built-up back pressure introduces a closing force on the unbalanced disc area. This force restricts application of these valves commonly to installations in which the valve discharges through a short pipeline that limits the developing built-up back pressure commonly to 10% of the set pressure. In the case of valves with a high built-in blowdown or which can be adjusted for a high blowdown, permissible built-up back pressures of 15% and higher are being quoted (consult manufacturer).

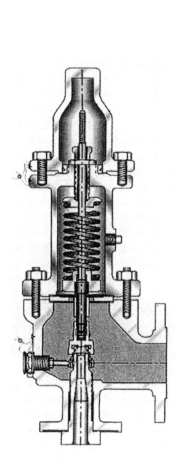

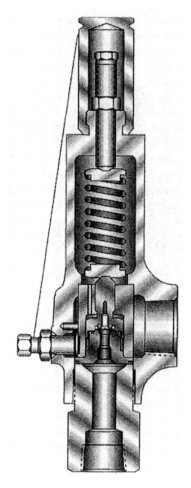

Figure 5-7. Conventional Safety Relief Valve. (*Courtesy of Anderson, Greenwood & Co.*)

Figure 5-8. Conventional Safety Relief Valve. (*Courtesy of Anderson, Greenwood & Co.*)

If the back pressure is superimposed and constant, the spring force can be adjusted for the differential pressure across the nozzle so that the valve may be employed for back pressures of up to 50% of the set pressure.

Figure 5-9 shows a safety relief valve that has been converted for steam duty by replacing the closed bonnet with an open bonnet as in safety valves. This modification converts the valve to a partially balanced pressure relief valve. As such, the valve may permit a built-up back pressure of 20%

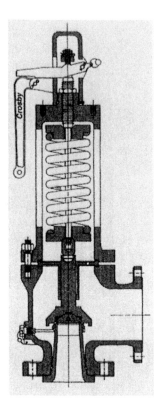

Figure 5-9. Safety Relief Valve with Open Bonnet. (*Courtesy of Crosby Valve & Gage Company, Wrentham, MA.*)

subject to confirmation by the valve manufacturer. The same conversion occurs also when a conventional safety relief valve with closed bonnet is provided with a regular or unpacked lifting lever used for hand lifting the valve disc from its seat.[1]

Balanced safety relief valves. Balanced safety relief valves such as those shown in Figure 5-10 through Figure 5-12 minimize the back pressure limitations of conventional safety relief valves by means of balanced bellows mounted between the valve disc and vented bonnet. By this means, the unbalanced disc area is exposed to constant atmospheric pressure. On valve opening, back pressure acts only on that portion of the disc in the closing direction that overhangs the valve seat. Although, back pressure acting on the convolutions of the bellows tends to elongate the bellows, manifesting itself in an increased spring rate of the bellows. As they are now stiffer, they resist valve lift accordingly.

---

[1] Refer to ASME Code Section VIII, Div. 1, UG-136 (3).

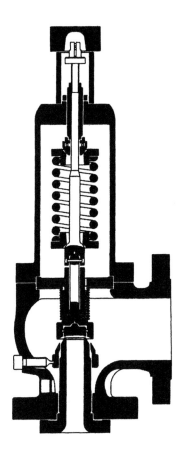

Figure 5-10. Bellows Balanced Safety Relief Valve, Bonnet Vented. (*Courtesy of Sempell A.G.*)

Balanced safety relief valves are quoted to achieve their rated capacity at back pressures between about 20% to 40%, depending on the design and size of valve. Then the valve begins to close as the back pressures rises. Allowing for reduced capacity, balanced safety relief valves are offered for back pressures of up to about 50% (consult manufacturer).

The bellows seal may be supplemented by an auxiliary mechanical seal as in the valve shown in Figure 5-11 to safeguard against bellows failure. Depending on the nature of the fluid handled by the valve, the vent may have to be directed toward a safe location.

Not all balanced safety relief valves offered in the market are truly balanced, though the balance achieved is quite acceptable. Possible exceptions are pressure relief valves to API Std. 526 of orifice sizes D and E, where the required diameter of the stem guide in relation to the nozzle diameter is

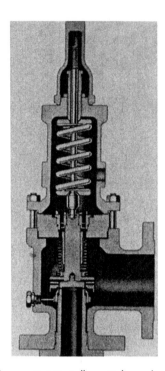

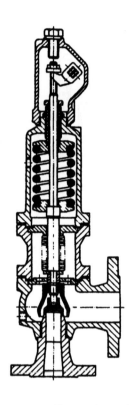

Figure 5-11. Bellows Balanced
Safety Relief Valve with Auxiliary
Piston Seal, Bonnet Vented. (*Courtesy
of Dresser Industries.*)

Figure 5-12. Bellows Balanced
Safety Relief Valve with Bellows
Mounted in Protective Spool, Bonnet
Vented. (*Courtesy of Bopp & Reuther.*)

too big to incorporate balanced bellows. Such valves are therefore affected
by variable back pressure. Some manufacturers restrict the lift of balanced
valves with F orifice to obtain an orifice D or E flow area.[45]

Blowdown adjustment of safety relief valves. Blowdown adjustment of
safety relief valves is commonly carried out by an adjustable nozzle ring
as found in the valve shown in Figure 5-7. Locating the adjustment ring on
the nozzle allows the disc guide to be located away from the direct impact
of the fluid escaping from the nozzle, leading to a robust construction for
process fluids that may carry solids in suspension. In combination with
balanced bellows, guide and valve spring can be completely isolated from
the flowing fluid.

The nozzle ring is used to control the width of the secondary orifice and hence also the pressure build-up in the huddling chamber. Raising the nozzle ring lengthens the blowdown but simultaneously reduces the operating pressure difference, resulting in a lowering of the popping pressure. Conversely, lowering the nozzle ring shortens the blowdown but simultaneously raises the popping pressure and increases simmer. Adjustment of the blowdown is therefore a compromise and must be carried out judicially. Usually manufacturers set the nozzle ring close to the disc holder to counter the possible adverse effect of inlet pressure loss on valve stability. The valve shown in Figure 5-12 omits the nozzle ring altogether.

In the valve shown in Figure 5-8, blowdown is adjusted externally by turning a screw, which partially restricts one of the holes in the disc guide. Adjustment of the screw varies the amount of developed back pressure below the disc that controls the blowdown without affecting the set pressure point. In conjunction with soft seatings, the valve as shown is offered for the thermal relief of small amounts of fluids such as LPG, ammonia, or other refrigerant type liquids where simmer between metal-to-metal seatings could cause severe icing.

Liquid Relief Valves

Normally, liquid relief valves differ from conventional safety relief valves only in a slight modification of the geometry around the disc to achieve a specific performance in liquid service. The valves shown in Figure 5-13 through Figure 5-15 represent a small selection of liquid relief valves offered by the industry.

The liquid relief valve shown in Figure 5-13 is designed to open fully within an overpressure of 10% and to offer stable operation over a wide range of operating conditions. Blowdown is adjustable by means of the nozzle ring.

Figure 5-14 shows a balanced liquid relief valve in which the stem carrying the disc is fitted with two O-ring seals. The lower O-ring seal serves as a drag ring that is energized by the inlet pressure through bore holes in disc and stem. The frictional resistance by this ring during opening and closing prevents uncontrolled chatter of the valve. The upper O-ring serves as a balanced seal against back pressure. A hole in the bonnet serves as a vent in case of leakage past the seal.

The valve shown in Figure 5-15 is a liquid relief valve with a linear proportional opening characteristic for operating conditions at which the relief valve is required to operate over a wide load range. By flaring the

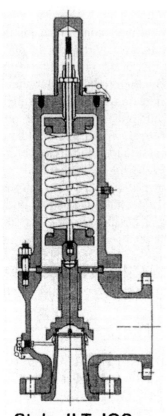

Style JLT-JOS

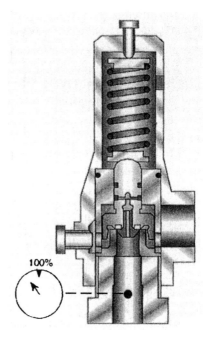

100%

Figure 5-13. Relief Valve. (*Courtesy of Crosby Valve & Gage Company, Wrentham, MA.*)

Figure 5-14. Balanced Relief Valve with Friction Ring to Prevent Valve Chatter. (*Courtesy of Anderson, Greenwood & Co.*)

nozzle outlet, changes in flow rate cause only small movements of the disc. Tests have shown that internal valve friction is sufficient to check valve oscillations even at rapidly accelerating mass flow. Figure 5-16 shows the opening and closing characteristics of the valve.

Effect of incompressible fluids on valve behavior. Because liquids are nearly incompressible and their density is high compared with gases, small changes in inlet flow velocity produce high pressure changes. During the opening and closing stages of the valve, the pressure changes translate immediately into changes of valve lift. The lift changes, in turn, influence

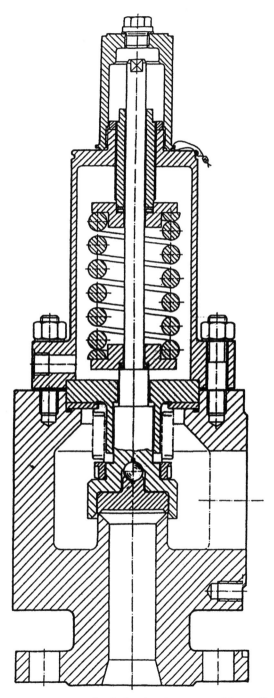

Figure 5-15. Proportional Relief Valve. (*Courtesy of Sempell A.G.*)

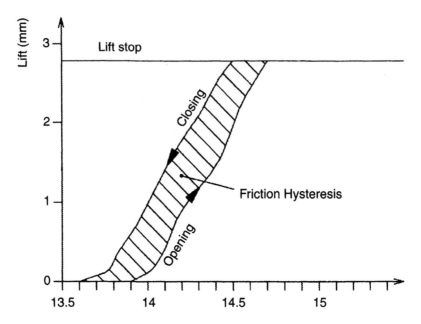

Figure 5-16. Opening and Closing Characteristic of Proportional Relief Valve shown in Figure 5-15. (*Courtesy of Sempell A.G.*)

the pressure changes. Liquid relief valves are therefore more readily prone to valve chatter than valves for gas service. However, fitting liquid relief valves with friction dampers as shown in the valve in Figure 5-14 and discussed in a subsequent chapter under the subject of "Oscillation Dampers" can avoid valve chatter altogether.

Vacuum Relief Valves

Figure 5-17 through Figure 5-19 show three types of direct-loaded vacuum relief valves that are or may be combined with positive over-pressure relief.

The valve shown in Figure 5-17 relies for loading solely on the light weight of the disc. To achieve a high degree of seat tightness under these loading conditions, the seat seal is made of sponge rubber. To also ensure easy travel of the disc in its guide, the guide rod is PTFE coated. The valve body is designed to be combined with an overpressure relief valve.

The pressure relief valve shown in Figure 5-18 combines a vacuum relief valve with a positive pressure relief. The valve has been designed for

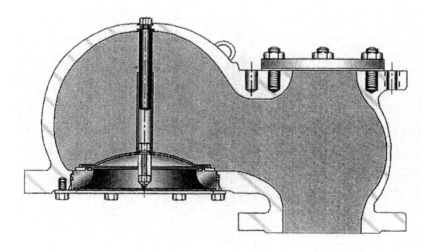

Figure 5-17. Direct-Loaded Vacuum Relief Valve. (*Courtesy of Anderson, Greenwood & Co.*)

sanitary application in the beverage, food processing, and pharmaceutical industries.

Figure 5-19 shows a breather valve that carries separately a direct-loaded vacuum relief valve and an overpressure relief valve. In deviation from other vacuum relief valve designs, the disc swings open in an arc on a point contact hinge. Valve seals consist of soft diaphragms that allow the valves to reseat close to the set pressure. The valve is intended for service on low-pressure storage tanks.

Direct-Loaded Pressure Relief Valves with Auxiliary Actuator

The need to improve the fluid tightness of spring-loaded pressure relief valves that have to operate close to the set pressure, and the need to inhibit valve chatter in difficult installations, led to the development of direct-loaded pressure relief valves that are combined with an auxiliary actuating device. The power to operate the auxiliary device is of an extraneous source, commonly compressed air in conjunction with electricity. Should the power supply to the auxiliary actuating device fail, the valve operates without interference as a spring-loaded pressure relief valve.

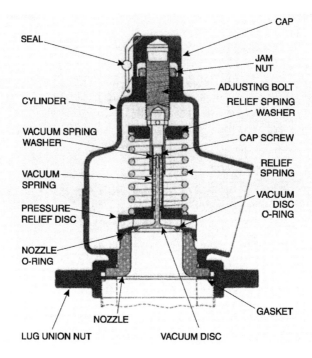

Figure 5-18. Direct-Loaded Vacuum/Pressure Relief Valve. (*Courtesy of Crosby Valve & Gage Co.*)

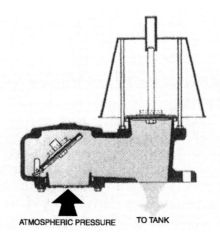

Figure 5-19. Breather Valve. (*Courtesy of Shand & Jurs.*)

Three types of spring-loaded pressure relief valves with auxiliary operating device have evolved over time to meet these needs:

- Assisted (anti-chatter) pressure relief valves[1]
- Supplementary-loaded (anti, simmer) pressure relief valves with restricted loading[2]
- Pressure relief valves with unrestricted supplementary loading and/or lift assistance[3]

Assisted pressure relief valves. Figure 5-20 shows the design of an assisted pressure relief valve and Figure 5-21 the associated control piping diagram. Under normal operating conditions, the assist device is deactivated. When the valve is called upon to open, the assist device is activated and introduces an unrestricted supplementary opening force. When the safe pressure is restored, the assist device is deactivated and the valve closes normally. If the external power supply should fail, the valve opens and closes with no interference from the assist device.

Supplementary loaded pressure relief valves with restricted loading. Figure 5-22 shows the design of this type of valve and Figure 5-23 the associated control piping diagram. Under normal operating conditions, the loading device introduces a supplementary but restricted closing force as permissible by the applicable Code of Practice. When the valve is called upon to open, the supplementary load is removed and the valve opens in the manner of a spring-loaded safety relief valve without interference by the loading device. Should the supplementary load fail to be removed, the valve will still fully open against the supplementary load to discharge its full rated capacity within the permissible overpressure.

This particular valve is fitted in addition with a mechanical fail-safe device for this mode of failure. By its design, the device disconnects the supplementary loading device from the valve stem as the overpressure rises. A lever fitted to the device permits the supplementary load to be released also by hand.

---

[1] ASME Code (Sect. III), European Standard EN 1268-1 (Safety Valves).
[2] ASME Code (Sect. III), European Standard EN 1268-1 (Safety Valves).
[3] European Standard EN 1268-5 (Controlled Safety Pressure Relief Systems) German Code of Practice AD Merkblatt A 2 and TRD 421.

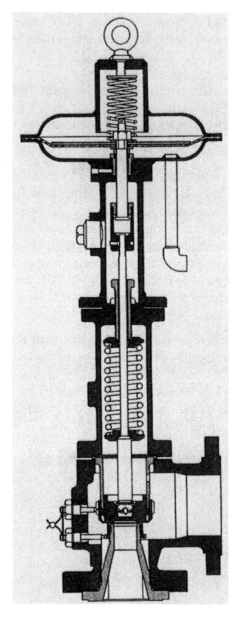

Figure 5-20. Assisted Safety Relief Valve. (*Courtesy of Crosby Valve & Gage Company, Wrentham, MA.*)

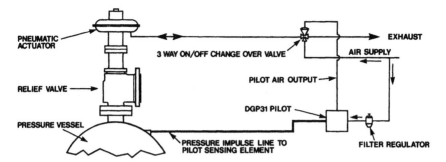

Figure 5-21. Typical Control Piping Schematic of Assisted Safety Relief Valve. (*Courtesy of Crosby Valve & Gage Company, Wrentham, MA.*)

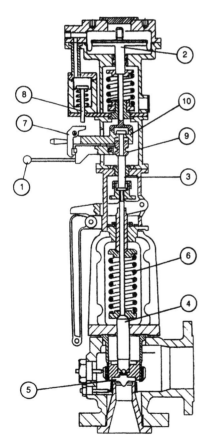

Figure 5-22. Supplementary Loaded Safety Relief Valve. (*Courtesy of Crosby Valve & Gage Company, Wrentham, MA.*)

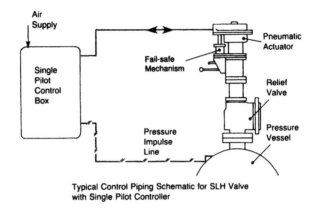

Typical Control Piping Schematic for SLH Valve
with Single Pilot Controller

Figure 5-23. Typical Control Piping Schematic of Supplementary-Loaded Safety Relief Valve. (*Courtesy of Crosby Valve & Gage Company, Wrentham, MA.*)

Should the power supply to the control device fail, the valve will open and close in the manner of a safety relief valve with no interference from the loading device.

Pressure relief valves with unrestricted supplementary loading and lift assistance. Figure 5-24 shows a spring-loaded pressure relief valve in combination with an auxiliary loading and lift-assist device represented by a pneumatic actuator. In this particular example, the control system shown incorporates three identical control lines. Using interlocked valving, one control line at a time may be taken out of service for testing, repair, or replacement without disturbing the operation of the pressure relief valve.

At normal operating conditions, the solenoid vent valves M1 to M3 are closed while the valve M4 closes against the compressed air system and vents the chamber H to the atmosphere. With the valves in these positions, compressed air entering the chamber B introduces a supplementary seating load.

When the pressure relief valve is called upon to open, the pressure switches deenergize the vent valves M1 to M3 that open to the atmosphere and energize the vent valve M4 that connects the compressed air system to the chamber H. With the valves in these positions, the chamber B is depressurized while the compressed air entering the chamber H assists valve opening.

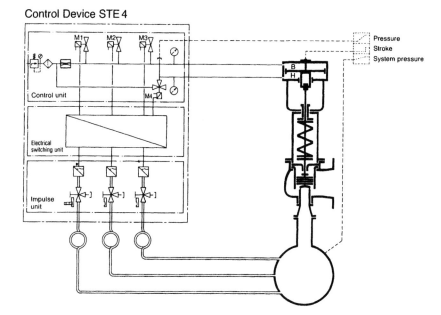

Figure 5-24. Schematic of Safety Relief Valve with Supplementary Unrestricted Loading and Lift Assistance with Associated Control System. (*Courtesy of Sempell A.G.*)

The control system permits close control of valve opening and closing. The system may also be adapted for either supplementary loading or lift assistance only.

Figure 5-25 and Figure 5-26 show two types of direct-loaded pressure valves for high pressure and high capacity duties in which the loading device consists of Belleville disc springs. In combination with an auxiliary actuator for supplementary loading and/or lift assistance and a control system identical or similar to the one shown in Figure 5-24, these valves are used in conventional and nuclear power plants.

Oscillation Dampers

Oscillation dampers are friction devices that act on the stem carrying the disc. Figure 5-14, Figure 5-27, and Figure 5-28 show three types of oscillation dampers in combination with direct-loaded pressure relief valves. The device shown in Figure 5-14 has already been described under the subject of "Relief Valves."

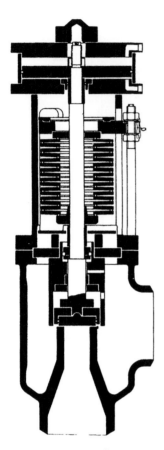

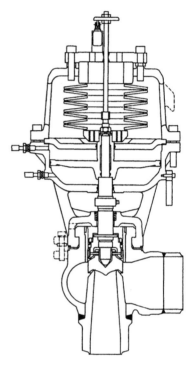

Figure 5-25. High Pressure, High-Capacity Safety Valve Combined with Auxiliary Pneumatic Actuator for Supplementary Loading and/or Lift Assistance. (*Courtesy of Sempell AG.*)

Figure 5-26. High Pressure, High-Capacity Safety Valve Combined with Auxiliary Pneumatic Actuator for Supplementary Loading and/or Lift Assistance. (*Courtesy of Bopp & Reuther.*)

The damping device of the pressure relief valve shown in Figure 5-27 relies on dry friction for damping oscillations, and uses lens-shaped friction elements made of hard electro graphite that are packed in a chamber between spring-loaded compression elements to force the friction elements against the stem. By this construction, the required amount of friction for a specific application can be controlled.

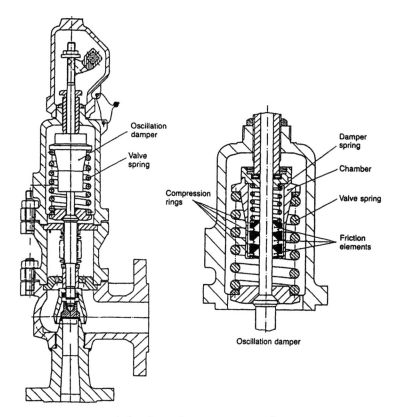

Figure 5-27. Pressure Relief Valve with Dry Friction Oscillation Damper. (*Courtesy of Bopp & Reuther.*)

Hard electro graphite has little dynamic friction scatter and the difference between static and dynamic friction is small. The friction force over the lift is therefore relatively uniform. Because static friction is low, its effect on the set pressure can be taken into account when setting the valve. Once the static friction has been overcome, there is no further impediment to the valve opening speed.

It is interesting to note that with very small amplitudes of oscillation, or at low frequencies, even a small amount of dry friction corresponds to a very large amount of equivalent viscous damping. This means that in the absence of significant disturbances, dry friction can be effective at preventing chatter. However, if the flow or pressure disturbances exceed

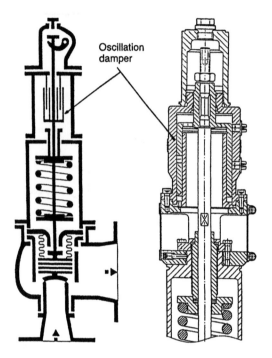

Oscillation damper

Figure 5-28. Pressure Relief Valve with Viscous Oscillation Damper. (*Courtesy of Sempell A.G.*)

the design limits of the dry friction device, the damper may not be able to damp out the oscillations.

Oscillation damping of the valve shown in Figure 5-28 relies on viscous friction. The damping device consists of close-fitting concentric cylinders with a narrow gap between them, which are filled with a special viscous grease. As the valve opens or closes, the cylinders glide relative to each other and introduce speed-dependent forces on the stem to damp out motion. The forces can be regulated by the width of the gaps between the cylinders and by the cylinder surface area.

Viscous damping adjusts itself to demand and is capable of preventing chatter regardless of the size of flow and pressure disturbance. The effect of viscous friction on the set pressure can be taken into account when setting the valve.

The combination of pressure relief valves and oscillation dampers as shown in Figure 5-27 and Figure 5-28 have been tested and certified by TüV, the technical supervisory organization in Germany.

Certification of Valve Performance

The ASME Boiler and Pressure Vessel Code requires pressure relief valve manufacturers to demonstrate performance and relieving capacity of pressure relief valves by a test to be carried out by an ASME-approved facility. The body that conducts the testing for certification of performance and capacity is the National Board of Boiler and Pressure Vessel Inspectors.[1] The methods and procedures to determine relieving capacities and additional characteristics that may be required by other codes are contained in the ASME Performance Test Code on Pressure Relief Devices PTC 25-1994 (latest revision).

The minimum design pressure of equipment covered by the ASME Code is for practical purposes 15 psi (1.03 bar). The ASME rules therefore do not generally apply to pressure relief valves for lower pressures.[2]

The certification tests should be conducted, with one exception, at an overpressure of either 3% or 10%, but minimum 3 psi (0.207 bar), as required by the applicable section of the code. Depending on the range of valves for which certification is sought, the following test methods are employed:

- **Three-valve Method** for capacity certification of one-valve model or type in one size at one set pressure.
- **Four-valve or Slope Method** for capacity certification of a type of valve of one size at various pressures.
- **Coefficient Method** for certification of the coefficient of discharge of a type or model of valve of various sizes over a range of pressures.

All values derived from the testing must fall within plus/minus 5% of the average value. This value is to be multiplied by 0.90 and the product be taken as the certified value. In the case of the coefficient of discharge, the average actual value is the K_d-value and the certified value the K-value. The derating factor of 0.90 compensates for minimal inlet pressure losses, for minus deviations from the average actual test value of up to 5%, wear and tear on the valve, and other unseen, unaccountable factors. The certified

---

[1] The National Board of Boiler and Pressure Vessel Inspectors, 1055 Crupper Avenue, Columbus, Ohio 43229-1183.

[2] For methods of capacity testing for conditions of subsonic nozzle flow refer to API Standard 2000.

valve capacities and coefficients of discharge are published by the National Board in the red book titled *Pressure Relief Device Certifications.*[1]

The blowdown of pressure relief valves with nonadjustable blowdown is to be noted and recorded as part of the performance test.

During all certification tests the valve must not flutter or chatter and shall properly reseat after the overpressure has receded.

Determining the permissible level of built-up and superimposed back pressure for satisfactory valve operation and its effect on mass flow or coefficient of discharge within the permissible back pressure range is not part of the performance test.

Force/Lift Diagrams as an Aid for Predicting the Operational Behavior of Spring-Loaded Pressure Relief Valves

Force/lift diagrams, such as those shown in Figure 5-29 to Figure 5-33, are important aids for predicting the operational behavior of pressure relief valves. They are also used for focusing solely on the effect of inlet pipe pressure loss and back pressure on the lifting force of the valve or on the mode of blowdown adjustment using a particular blowdown adjustment device.

The straight dashed line shown in the diagrams represents the closing force exerted by the spring on the disc. The S-shaped curves represent the net opening forces from the fluid acting on the disc. These forces are determined experimentally at constant inlet pressure. The actual shape of the curves varies with valve designs.

Decaying effect of inlet pressure loss on lifting force. The force/lift diagram (Figure 5-29) shows the decaying effect of rising inlet pipe pressure loss on the lifting force, as taken for conditions of zero back pressure. The type of valve used in this test is similar to that shown in Figure 5-5. In this particular case, the lift has been restricted to $l/d_0 = 0.22$.

---

[1] The coefficients of discharge contained in the red book of the National Board are correlated to the actual valve flow area. This area may be larger than the standard "effective" orifice area specified in the API Standard 526. Where the "effective" orifice area is employed in the sizing equations as in API RP 520, the correlated "effective" coefficient of discharge must be used in sizing the valve as derived from the relationship $(AK)_{Certified} = (AK)_{Effective}$.

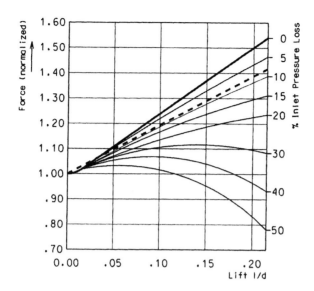

Figure 5-29. Force/Lift Diagram Displaying the Decaying Effect of Inlet Line Pressure Loss on the Lifting Force of a Pressure Relief Valve at Zero Back Pressure. (*Courtesy of Sempell A.G.*)

Based on this diagram, the manufacturer has chosen a spring characteristic, which causes the valve to open fully in one stroke at an inlet pressure loss of 3%.

Decaying effect of built-up back pressure on lifting force. The force/lift diagram (Figure 5-30) shows the decaying effect of built-up back pressure on the lifting force of an unbalanced conventional pressure relief valve similar to that shown in Figure 5-5, as taken for conditions of zero inlet pipe pressure loss. To determine the lifting force for conditions of 3% inlet pipe pressure loss and a given built-up back pressure, the appropriate lifting force curves shown in Figure 5-29 and Figure 5-30 must be superimposed. If the built-up back pressure at 3% inlet pipe pressure loss is high enough, the valve will lose its ability to open fully in one stroke.

Modes of blowdown adjustment. The force/lift diagrams (Figure 5-31 (a) and (b)) display the mode of blowdown adjustment with devices that are typical of pressure relief valves shown in Figure 5-5 and Figure 5-7.

Pressure Relief Valves

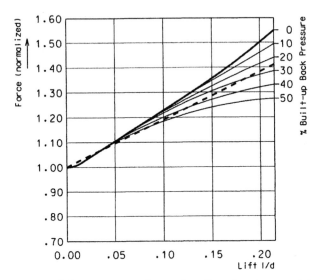

Figure 5-30. Force/Lift Diagram Displaying the Effect of Built-Up Back Pressure on the Lifting Force of an Unbalanced Pressure Relief Valve at Zero Inlet Line Pressure Loss. (*Courtesy of Sempell A.G.*)

The force/lift diagram shown in Figure 5-31 (a) applies to pressure relief valves of the type shown in Figure 5-5. The disc of the valve is guided in a sleeve, which is provided with a screwed ring or sleeve extension. Screwing down the ring extends the overall sleeve length, and thereby raises the angle of deflection of the fluid acting on the underside of the disc. The resulting higher lifting force, in turn, lengthens the blowdown. Because the width of the circumferential orifice between seat and guide sleeve remains constant during valve opening, lowering the guide ring causes the lifting force to rise nearly in proportion to the rise in valve lift. Lengthening the blowdown, however, lowers simultaneously the popping pressure by some amount.

The valve is fitted in this case also with a nozzle ring for blowdown adjustment. This ring is set by the manufacturers and should not normally be adjusted by the user.

The force/lift diagram shown in Figure 5-31 (b) applies to pressure relief valves of the type shown in Figure 5-7. The blowdown adjustment device consists in this case of a screwed ring that is mounted on the valve

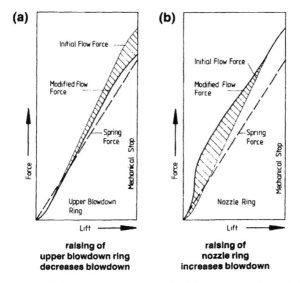

Figure 5-31. Force/Lift Diagram Displaying the Mode of Blowdown Adjustment Using Two Different Types of Blowdown Adjustment Devices. (*Courtesy of Sempell A.G.*)

nozzle. Raising the adjustment ring towards the valve disc reduces the width of the secondary orifice around the seat. On receding over pressure, the reduced width of the secondary orifice will cause the disc to hover above the seat until the system pressure has dropped by some amount. The delayed reseating of the disc causes the blowdown to lengthen. Raising the blowdown, however, raises the popping pressure by some amount.

Operational behavior of spring-loaded pressure relief valves that pop open fully in one stroke (typical only). The force/lift diagram (Figure 5-32), displays the relationship between lift force and valve lift over the entire operating cycle of a pressure relief valve. Inlet pipe pressure loss is assumed to be 3%. Built-up back pressure in discharge pipe is undefined.

When rising system pressure reaches the set pressure at operating point A, the valve will modulate open until reaching the operating point B. At this point, lifting and closing forces are in balance. At the commencement of any further pressure rise, the lifting force grows faster than the spring force, causing the valve to pop open fully in one stroke to operating

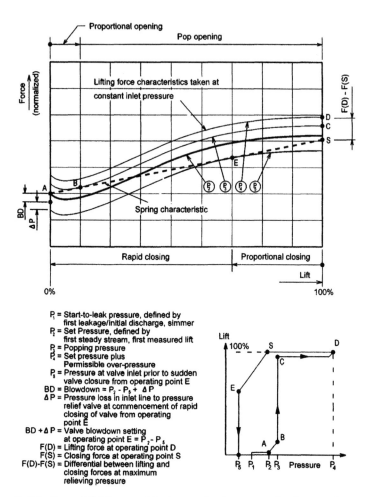

Figure 5-32. Force/Lift Diagram Displaying the Operational Behavior of Conventional Spring-Loaded Pressure Relief Valves that Pop Open Fully in One Stroke (Typical Only).

point C. As the system pressure rises further as permitted by the applicable code of application, the valve reaches its rated capacity at operating point D.

On receding over pressure, the valve retains full life until reaching the operating point S, where lifting and closing forces are again in balance. As the system pressure recedes further, the valve modulates closed until reaching the operating point E.

At this point, the difference between the set pressure and the pressure at the valve inlet is the sum of the pressure loss in the inlet line to the valve and the blowdown at valve reseating. This pressure difference is referred to in the following as the blowdown setting of the valve. As the operating pressure commences to recede further, the closing force begins to exceed the lifting force, causing the valve to reseat in one rapid stroke while the pressure loss in the inlet line to the valve drops progressively to zero.

If the blowdown setting of the valve at operating point E were less than the pressure loss in the inlet line to the valve, the valve could not close completely. The valve would then reopen and reclose in rapid succession. Acoustic interaction between pipeline and valve could severely aggravate valve chatter. Chatter would continue until the operating pressure has dropped sufficiently or the valve has destroyed itself.

The diagram below the force lift diagram shows the corresponding relationships between valve lift and operating pressure. The diagram, referred to as the lift/pressure diagram, shows the periods of rapid and modulating movements of the disc over the operating cycle from the set pressure to the reseating pressure.

Operational behavior of spring-loaded pressure relief valves that pop open partially only and then continue to modulate open (typical only). The force lift diagram shown in Figure 5-33 is based on the lift/pressure relationship shown in Figure 22 of API RP 520 Part I (1993). Valves with this operating characteristic pop open over a portion of the full lift only and then continue to modulate open until the full lift is achieved. As the operating pressure rises still further, the valve achieves its rated capacity at operating point D. This opening characteristic is common to a wide range of spring-loaded pressure relief valves.

The difference between the lifting force at operating point D and the closing force at operating point S is a measure of the tolerance of a particular valve to back pressure. If the force/lift diagrams (Figure 5-32 and Figure 5-33) apply to different types of valves but identical operating conditions, the valve with a force/lift characteristic of Figure 5-32 will display a higher resistance to back pressure than the valve with a force/lift characteristic of Figure 5-33. The back pressure tolerance of the latter valve can be improved within limits by lowering the incline of the spring characteristic. This method of improving the back pressure tolerance of a pressure relief valve is offered by a number of

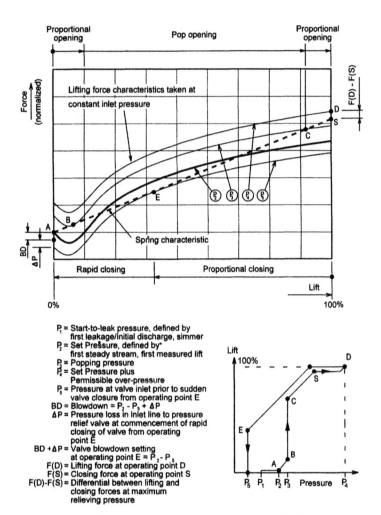

Figure 5-33. Force/Lift Diagram Displaying the Operational Behavior of Conventional Spring-Loaded Pressure Relief Valves that Pop Open Partially Only and Then Continue to Modulate Open (Typical Only).

manufacturers. Alternatively, choose a valve with the next larger size orifice but restrict the valve lift to the required flowrate. While the higher back pressure tolerance is achieved by shortening the decaying end of the lifting force curve, larger inlet and discharge piping of the valve also lower the inlet pressure loss and built-up back pressure from flow through the valve.

Secondary Back Pressure from Flow-Through Valve Body

A secondary back pressure develops from the flow-through valve body between valve seat and valve outlet flange. For sonic flow in the valve outlet branch, this back pressure is intrinsic to the valve.[46]

The back pressure arising from this flow causes the lifting force to decay in the same manner as back pressure that is present in the outlet pipe. If the secondary back pressure is high enough, the valve may not be able to achieve its rated capacity.

The back pressure that can develop in the valve body is related to the area ratio of valve outlet size to nozzle size. For pressure relief valves of API Std. 526, this area ratio varies considerably between valve sizes as shown in Table 5-1. For example, the area ratio of valve size $1\frac{1}{2}$ D $2\frac{1}{2}$ is about 14 times larger than the area ratio of the valve 8T10. A conventional safety relief valve of size 8T10 used on steam has been reported to not achieve its rated capacity at a set higher than 50 psig (3.5 barg), even with no outlet piping. At this pressure level, the size of the valve ought to be 8T12. Thus, when applying pressure relief valves with a low area ratio of valve outlet size to nozzle size, caution should be applied. Consult valve manufacturer when in doubt.

Recommended maximum non-recoverable pressure loss in piping between protected equipment and inlet of direct-acting pressure relief valves.[1,2] The following recommendation has been universally accepted:

- The non-recoverable pressure loss in the pipeline between the protected equipment and the direct-acting pressure relief valves should not exceed 3% of the set pressure of the valve.
- When the pressure relief valve is installed on a process line, the 3% limit should be applied to the loss in the normally nonflowing pressure relief valve inlet pipe and the incremental pressure loss in the process line caused by the flow through the pressure relief valve.

---

[1] Refer to API RP 520 Part II. Fourth Edition, paragraph 2.2.2.
[2] For permissible pressure loss in inlet line to pilot-operated pressure relief valves with direct-acting pilot, refer to "Stable operation of valves with on/off pilots".

Table 5-1
Area ratio of valve outlet size to nozzle size of pressure relief valves to API Std. 526[47]

Inlet	Outlet	Orifice	Area ratio Outlet/Orifice
$1,1\frac{1}{2}$	2	D	30.55
$1\frac{1}{2}$	$2\frac{1}{2}$	D	43.55
$1,1\frac{1}{2}$	2	E	17.14
$1\frac{1}{2}$	$2\frac{1}{2}$	E	24.44
$1\frac{1}{2}$	2	F	10.94
$1\frac{1}{2}$	$2\frac{1}{2}$	F	15.60
$1\frac{1}{2}$	$2\frac{1}{2}$	G	9.52
2	3	G	14.71
$1\frac{1}{2}$	3	H	9.43
2	3	H	9.43
2	3	J	5.75
$2\frac{1}{2}$	4	J	9.91
3	4	J	9.91
3	4	K	6.94
3	6	K	15.73
3	4	L	4.47
4	6	L	10.14
4	6	M	8.03
4	6	N	6.66
4	6	P	4.53
6	8	Q	4.52
6	8	R	3.12
6	10	R	4.93
8	10	T	3.03

The rated capacity of the valve at 10% over pressure for determining the pressure loss in the inlet pipeline is equal to the actual or non-derated valve capacity of the fully open valve at zero overpressure. The amount of inlet pressure loss that is part of the valve blowdown setting value depends on the lift of the valve at which closure occurs in one rapid stroke. If valve closure in one rapid stroke occurs from the fully open position, the pressure loss component of the valve blowdown setting is a maximum and should normally not exceed 3% of the set pressure. Should the valve be capable of modulating from the fully open to the fully closed position, the pressure loss component of the valve blowdown setting would be zero. In most installations, however, rapid valve closure occurs from the partly open position.

• In both cases, the pressure loss should be calculated using the rated capacity of the valve at an overpressure of 10% (see footnote on page 210).

Keeping the pressure loss below 3% becomes progressively more difficult as the value of the area ratio of valve entry to valve nozzle A_E/A_N decreases.[47] For a ratio of 0.9 $A_E/KA_N = 4$ for a full lift liquid relief valve, for example, the value of the permissible resistance coefficient is only about 0.5 (see sizing diagram Figure 7-8 on page 285). This is already the resistance coefficient of a sharp-edged pipe inlet. In this particular case, either a larger inlet pipe should be considered or, where this is not possible, a liquid relief valve with oscillation damper. Consult manufacturer in the latter case.

Instability due to valve oversizing. If a pop action pressure relief valve is grossly oversized, the discharging fluid may not be able to hold the valve in the open position. The valve will then reopen and close in succession. According to one manufacturer's catalog,[48] this may happen if the rate of flow is less than approximately 25% of the valve capacity.

The frequency of valve opening and closing for a given mass flow depends on the volume of the system and the valve blowdown setting. Instability can be overcome entirely by limiting the valve lift in consultation with the valve manufacturer.

Verification of Operating Data of Spring-Loaded Pressure Relief Valves Prior to and After Installation

Verification of operating data of pressure relief valves prior to installation consists commonly of a set pressure test only. The test is normally carried out on a test bench as instructed by the valve manufacturer. Under bench conditions, the set pressure point can readily be registered by observing first steady flow or lift.

However, only verification testing *in situ* can register the plant specific influences on set pressure and operation of the valve. But because plant noise interferes with the audible detection of the set pressure, the popping pressure is commonly taken as the set pressure. This assessment method is of little consequence for high pressure valves in which the opening pressure difference amounts often to not more than 1% to 2% of the set pressure. In low pressure valves, however, the popping pressure may be higher than the actual set pressure by as much as 5%. This deviation must be taken into account when correlating the thus determined set pressure to the normal operating pressure, the reseating pressure, and the assessment of the achieved blowdown. Thus, if in addition, the operational behavior of the valve has to be assessed, the plant operating pressure has to be raised further by about 5% or 10%.

To overcome the need for overpressurizing the plant, portable verification devices such as those shown in Figure 5-34 have been developed that permit initial and periodic *in-situ* testing of pressure relief valves without interfering with normal plant operation. These devices simulate overpressure by introducing a measurable lifting force on the valve stem that may be produced either mechanically, hydraulically, or pneumatically as in the device shown in 5-34. There is no need for controlling the system pressure.

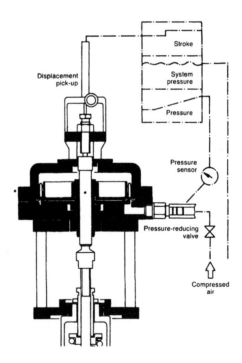

Figure 5-34. Portable Device for the *In-Situ* Verification of Valve Operating Data without Interfering with Plant Operation. (*Courtesy of Sempell A.G.*)

The duration of the actual tests is short and measured in seconds rather than minutes. The values that may be verified are:

- Set pressure
- Popping pressure
- Blowdown of reseated valve
- Valve lift
- Spring rate

Part of the equipment is a lap top computer with special software and a printer. This equipment delivers the measured data and a pressure/lift diagram. The output allows a complete performance evaluation of the tested valve.

The set pressure may be verified at any system pressure and at the test bench. The other data may be evaluated successfully only at an operating pressure of not less than 80% of the set pressure.

PILOT-OPERATED PRESSURE RELIEF VALVES

Pilot-operated pressure relief valves differ from direct mechanically loaded pressure relief valves in that the system fluid is the medium for both opening and closing the valve. Such valves consist of a main valve, which is the actual pressure relief valve, and a self-actuated pilot that controls the opening and closing of the main valve in response to developing and receding overpressure.

Over the years, two types of pilot-operated pressure relief valves have evolved.

One of these is of U.S. origin in which the pilot controls directly three flow passages of which one interconnects with the pressure sense line, a second one with the dome of the main valve, while a third one represents the vent. This type of pilot is described in the following as direct-acting pilot.

The other type is of European origin in which the pilot represents a spring-loaded pressure relief valve that controls the operation of the main valve by the fluid being discharged.[1] This type of pilot is described in the following as indirect-acting pilot.

---

[1] European Standard EN 1268-1.

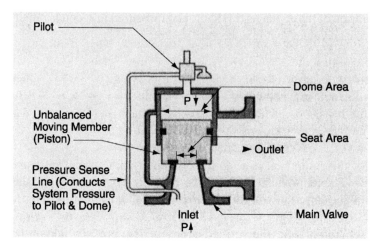

Figure 5-35. Basic Layout of Pilot-Operated Pressure Relief Valve with Direct-Acting Pilot. (*Courtesy of Anderson, Greenwood & Co.*)

Pilot-Operated Pressure Relief Valves with Direct-Acting Pilot

Figure 5-35 shows the basic layout of a pilot-operated pressure relief valve with direct-acting pilot. The actual pressure relief valve is the main valve in which the piston is the closing member that is controlled by the pilot.

At a system pressure lower than the set pressure, the system pressure is allowed to act on the top of the main valve piston while the vent to the atmosphere is closed. Because the piston area facing the dome is larger than the nozzle area, the system pressure introduces a closing force on the piston that grows with rising system pressure. When the system pressure reaches the set pressure, the pilot vents the dome pressure sufficiently to allow the pressure on the underside of the piston to force the valve open. When receding overpressure reaches the predetermined closing pressure, the pilot closes the vent and permits the system pressure to act on the main valve piston in the closing direction.

Pilot types. Pilots may be of the flowing or non-flowing type.

Flowing type pilots permit the system fluid to continue to bleed through the pilot to vent during the entire venting cycle.

Non-flowing pilots, on the other hand, stop the bleed of system fluid into the dome volume when called upon to vent the dome. Stopping the bleed of system fluid into the dome during the valve-opening cycle minimizes the possibility of impurities entering the pilot that could interfere with the pilot operation. In general, the non-flowing pilot is the preferred pilot.

Opening and closing actions of pilots. Pilots may be designed for either snap on/off or modulating opening and closing action.

On/off action pilots introduce the full system pressure to the dome until the set pressure has been reached. Then the pilot vents the dome pressure, causing the main valve to fully pop open. When the system pressure has dropped to the blowdown pressure, the pilot reintroduces the system pressure to the dome, causing the valve to drop closed.

Modulating pilots are designed to open and close in increments according to demand so that mass flow through the main valve is the same as at the pressure source into the system. Such pilots introduce the full system pressure to the dome until the dome pressure commences to decay at a system pressure of about 95% of the set pressure. When the set pressure has been reached, the valve commences to open and can be fully open at an overpressure of about 105% of the set pressure. On receding overpressure, the closing motion is in reverse.

Balanced and unbalanced pilots. Like spring-loaded pressure relief valve, pilots may be back pressure balanced or unbalanced.

In the case of balanced pilots, the dome volume may be vented directly to the main valve outlet as long as the back pressure does not exceed the system pressure.

In the case of unbalanced pilots, however, the dome must be vented to the atmosphere at the pilot or to a preferred location to ensure proper functioning of the pilot. Within this one constraint, the set pressure of pilot-operated pressure relief valves is unaffected by back pressure in the main valve outlet unless the situation arises in which the back pressure exceeds the inlet pressure. To prevent reverse flow occuring in this case, manufacturers provide the pilot on request with an appropriate check valve arrangement, referred to as backflow preventer.

Types of main valves. Main valves are commonly classified by type as diaphragm type, bellows type or piston type, depending on the type of moving member that opens and closes the valve.

Figure 5-37 depicts a diaphragm type and Figure 5-39 a piston type main valve in conjunction with a pilot. Diaphragm type main valves are used on low pressure services only. Should the valve discharge into a vessel in which unforeseen unwanted vacuum could develop, the diaphragm would lift off its seat and protect the vessel from collapsing.

Direction of flow through main valves. The direction of flow through main valves is commonly with the fluid flow acting on the underside of the closure member, as shown in Figure 5-35.

The valve shown in Figure 5-41 deviates from this practice by reversing the direction of flow through the valve. By this flow reversal, piston and guide are located in the pressure zone of the valve. This location of piston and guide has specific advantages in high pressure and high temperature applications. First, piston and guide are at the temperature of the system fluid and therefore do not suffer thermal shock on valve opening. Second, solids carried by the discharging fluid do not impact directly on the seating surface of the piston. Location of the piston within the pressure zone therefore significantly reduces potential seating damage. The opening and closing deceleration of the piston is controlled in this case by built-in dampers to limit mechanical shock on valve and piping system on valve opening and closing.

Standard and full-bore flanged steel main valves. API Standard 526, which covers end connections, nozzle sizes, and pressure and temperature limits of flanged steel spring-loaded pressure relief valves, has been extended to cover pilot-operated valves as well as spring-loaded pressure relief valves.

As a carryover of the time when above standard did not apply to pilot-operated valves, most U.S. manufacturers also offer non-standard pilot-operated pressure relief valves, referred to as full-bore valves. The term full-bore applies to the nozzle size, which approaches the size of the bore of the inlet pipe.

Figure 5-36 shows an example of a full-bore valve. It is noteworthy in this case that some bodies can accommodate a number of orifice sizes by adjusting the length of the piston stroke by means of an adjustable length bolt. The orifices involved cover sizes D through T according to API 526.

The piston is provided in this presentation with a pressure-actuated drag ring designed to suppress rapid oscillating movements of the piston due to resonant reinforcement of the assembly and inlet piping.

Illustrations of pilot-operated pressure relief valves. Figures 5-36 through 5-41 show a cross section of the diverse range of pilot operated pressure relief valves with direct-acting pilot, covering the

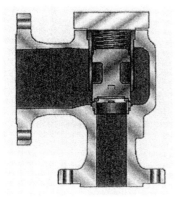

Figure 5-36. Full-Bore Flanged Steel Main Valve. (*Courtesy of Anderson, Greenwood & Co.*)

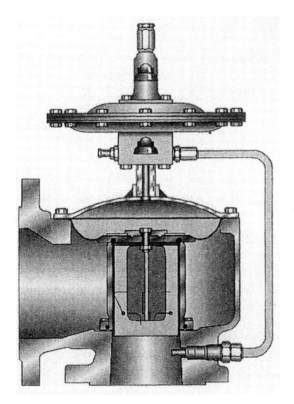

Figure 5-37. Pilot-Operated Pressure Relief Valve, Diaphragm Type, Unbalanced, Flowing Pilot, Snap or Modulating Action, May Be Piped-up for either Overpressure or Vacuum Relief. Alternative Pilot: Balanced, Non-Flowing, Modulating Action. Application: Low Pressure Gas. (*Courtesy of Anderson, Greenwood & Co.*)

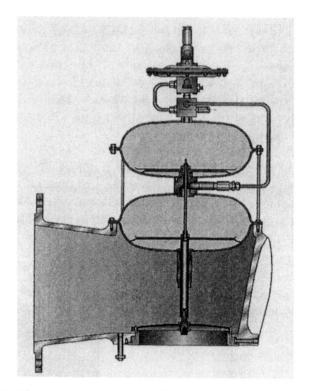

Figure 5-38. Pilot-Operated Pressure Relief Valve, Diaphragm Type for Overpressure and Vacuum Relief Modes, Balanced Non-Flowing Pilot, Modulating Action. Application: Low Pressure Gas. (*Courtesy of Anderson, Greenwood & Co.*)

wide field from vacuum relief to high pressure and high temperature applications.

Accessories. Pilot-operated pressure relief valves may be provided with additional accessories to provide additional functions. Others may be used to assist in the successful operation of the pressure relief valve. The following include a range of such accessories:

- Field test connection for the in-service verification of the set pressure.
- Field test indicator for the in-service verification of set pressure of modulating type pilots only.
- Backflow preventer to prevent accidental reverse flow through the valve if outlet pressure ever exceeds inlet pressure.
- Pilot supply filter to protect the pilot from entry of significant impurities carried by the fluid traveling in the pressure sense line. In the case of

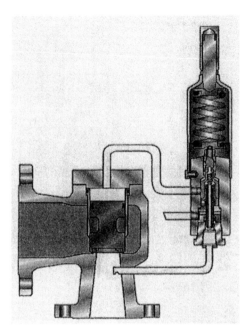

Figure 5-39. Pilot-Operated Pressure Relief Valve, Pilot-Balanced, Non-Flowing Modulating Action. Application: Gas, Liquid and Two-Phase Fluids. (*Courtesy of Anderson, Greenwood & Co.*)

liquid service, the filter is particularly appropriate in connection with flowing type modulating pilots.

• Pressure spike snubber to attenuate pressure spikes in gas systems downstream of reciprocating compressors.

• Manual or remote unloader permits the main valve to be opened remotely to depressurize the system.

• Pilot lift lever to permit manual testing of the pressure relief valve operation. (Note: Lift lever does not permit checking the set pressure as sometimes assumed.)

• NACE material option for sour gas service in accordance with the requirements of NACE MR0175 for both pilot and main valve.

Stable Operation of Valves with On/Off Pilots

These are the rules:

The pressure loss in the valve inlet pipeline section between the integral sense pressure pick-up point and the source of pressure, as with spring-loaded pressure relief valves, must be less than the pilot blowdown. If this

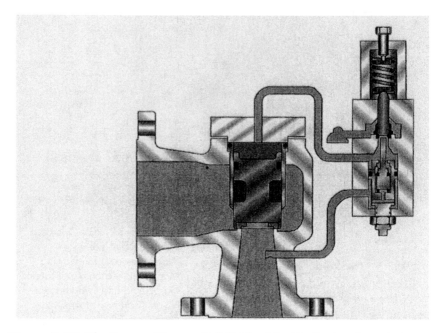

Figure 5-40. Pilot-Operated Pressure Relief Valve, Pilot Nonflowing, Snap Action Type. Application Primarily for Gas, in Particular, Steam. This Includes Severe Service Where Dirt, Hydrates, and High Moisture Levels Occur in the Fluid Medium. Set Pressures 25 to over 6000 psig (1.7 to over 400 barg), Continuous Service Temperature −423°F to +500°F (−253°C to 260°C). (*Courtesy of Anderson, Greenwood & Co.*)

is not observed, the valve will reclose after opening and then reopen again in rapid succession until the valve has destroyed itself or the system pressure has risen sufficiently to stop valve cycling. It is prudent to ensure that the blowdown is at least 2% longer than the pressure loss in the pipe section upstream of the sense pressure pick-up point. Note that modulating pilots overcome the potential for cycling in this situation.

The length of remote pressure sense lines is determined by the distance between the location of the pressure relief valve and the pressure sensing point. Such lines must be run in a manner that prevents condensate from blocking the flow passage. In the case of pressure relief valves with non-flowing pilots, pressure sense lines of 200 feet (60 m) length have been reported, using a size 1"NB (DN 25) pressure sense line in compliance with the applicable piping code for the system. In the case of steam service and other services with the potential of high condensate flow, the valve should be mounted as close as possible to the pressure source.

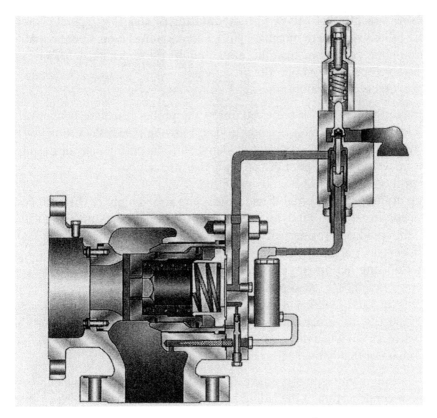

Figure 5-41. Pilot-Operated Pressure Relief Valve, Pilot Nonflowing Type, Snap Action. Optional Pilot for Air, Gas, and Vapor Only, or for Saturated and Superheated Steam, Maximum Set Pressure 1200 psig (82.8 barg), Maximum Temperature 1000°F (538°C), Adjustable Blowdown, Remote Actuation and Dual Pilot Option with Integral Selector. (*Courtesy of Anderson, Greenwood & Co.*)

If the piston of the main valve is undamped, flow disturbances in the inlet pipeline such as caused by a single elbow or an equivalent restriction can already lead to valve chatter.[49] In the case of full-bore main valves, stability may be achieved by limiting the length of the inlet pipeline to an L/D of 5. This requires an excellent rounded entry to the pipeline. Where longer pipe runs are required, the inlet pipe size should be one size larger than the valve inlet and be used with a concentric reducer below the valve. In less than ideal installations, frictional resistance to the piston movement can completely suppress valve resonant chatter. This has been achieved in the valve shown in Figure 5-36 by a pressure-activated drag device between piston and cylinder.

Stable operation of valves with modulating pilots. Pilot-operated pressure relief valves with modulating pilot are more forgiving to unfavorable installation conditions than those with on/off pilot, primarily because of the valve opening in proportion to demand.

Several types of pilots are available:

Flowing type: (a) Not preferred unless the process medium is relatively clean. (b) Preferred for steam service. Flowing type allows the pilot to warm up internally to prevent sudden steam condensation inside the pilot that would lead to pilot instability and erratic action.

Non-flowing pilot is generally preferred.

Pop action: (a) Preferred if rate of pressure rise is high, or (b) if process medium can form solids after expanding through the pressure relief valve that could plug or partially plug the discharge header. Example: CO_2 forms dry ice particulate upon expansion.

Modulating action: (a) Preferred when required discharge capacity can vary greatly; (b) also preferred for operation in liquid service; (c) to ensure valve stability if the valve is oversized or, within limits, undersized due to underestimating the inlet pressure loss when sizing the valve, or (d) when the state of process phase can vary. Example: Oil/gas separator that is liquid-flooded.

Consult manufacturer if fluid carries solids in suspension or when freezing or condensation of the lading fluid at ambient temperature is possible. Do not use pilot-operated valves in abrasive or viscous services (beyond 5000 SSU) and in services in which coking, polymerization or corrosion of the wetted parts of pilot and main valve can occur, unless the valve is designed for this purpose. Pilot-operated pressure relief valves used for this purpose are speciality valves in which the critical internals of main valve and pilot are protected from the process medium by a clean external medium, even though the pilot is actuated by the process fluid.

Pilot-Operated Pressure Relief Valves with Indirect-Acting Pilot

Pilot-operated pressure relief valves with indirect acting pilots may be of two principal types, one in which the main valve is being deenergized to open and one in which the main valve is being energized to open.

The pilot shown in Figure 5-42b is a spring-loaded safety valve designed in this case for steam. The main valve shown is of the deenergize-to-open

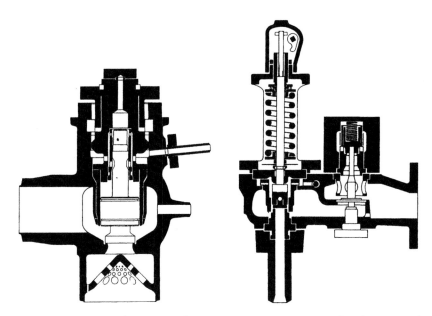

Figure 5-42. Main Valve Designed to Open on Being Deenergized and Associated Direct-Acting Pilot. (*Courtesy of Sempell A.G.*)

type. When the pilot opens on rising overpressure, the discharging fluid acts on a check valve, which in turn vents the dome of the main valve shown in Figure 5-42a. The system pressure acting on the unbalanced area of the main valve piston causes the main valve to open. While the valve is open, a continuous small bleed into the dome above the piston is maintained. When on receding overpressure the pilot closes, the check valve returns to its closed position and closes the dome vent to the atmosphere. The bleed entering the dome now closes the valve. The perforated cone in the outlet connection of the main valve serves as a diffuser type silencer.

Figure 5-43 shows an installation diagram of the main valve in conjunction with three control lines, two of them carrying indirect-acting pilots as shown in Figure 5-42 and the third a solenoid control valve. The isolation valves on each side of the pilots and the solenoid control valve are interlocked in a manner that allows one control line at a time to be isolated for inspection and maintenance. The number of control lines is by choice of the user or as required by the code of practice.

The main valve shown in Figure 5-44 is of the energize-to-open type. The closure member is a cylinder-shaped vessel that moves on a central piston. The pilot is a bellows sealed direct spring-loaded pressure relief valve that

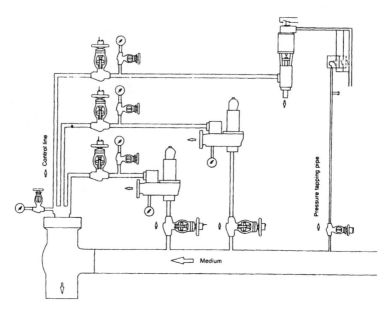

Figure 5-43. Installation Diagram of Main Vale with Indirect Acting Pilots, Designed to Open on Being Deenergized (Typical Only). (*Courtesy of Sempell A.G.*)

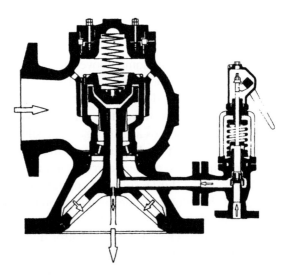

Figure 5-44. Main Valve of a Pilot-Operated Pressure Relief Valve with Indirect Acting Pilot, Designed to Open on Being Energized. (*Courtesy of Bopp & Reuther.*)

may be provided with supplementary loading and assisted lifting by means of solenoids.

When the system pressure rises and reaches the set pressure, the pilot opens and introduces the system pressure via a central tube to the top of the piston. The pressure build-up above the piston causes the main valve to open. Some of the pressure is allowed to bleed through a small hole to the valve outlet.

When the excess pressure recedes and reaches the blowdown pressure, the pilot closes and allows the pressure above the central piston to bleed to the valve outlet. A spring above the closing member assists the valve to close suddenly.

In both types of main valves, the closure members are located within the system pressure zone and are exposed to the temperature of the system pressure. In the case of high temperature service, the closure member and guide are therefore not exposed to thermal shock. Also, solids carried by the flowing fluid do not impact directly on the seating faces of the closure member. This flow direction significantly reduces potential seating damage.

Both types of pilot-operated pressure relief valve are widely employed in conventional and nuclear power stations.

6

RUPTURE DISCS

The use of a rupture disc is the most basic method of protecting a piping system from overpressurization. A rupture disc is a sacrificial component and after the disc has been ruptured during overpressurization, this component must be replaced to protect the vessel and the piping system.

Rupture discs are the pressure and temperature sensitive element of non-reclosing pressure relief devices, consisting of the rupture disc and a holder. They are designed to protect pressure systems against damage from excessive overpressure or vacuum by bursting at a predetermined pressure differential across the disc.

The original rupture disc consisted of a plain metal sheet that was clamped between two flanges. When exposed to pressure on one side, the disc would stretch and form a hemispherical dome before bursting. The predictability of the burst pressure, however, was poor. To improve the predictability, rupture discs were subsequently predomed by applying pressure to one side of the disc that was higher than the normal operating pressure by some margin.

The rupture disc thus produced is today's solid-metal forward-domed rupture disc. Flat metal rupture discs have also been reengineered for low-pressure applications. Both types of rupture discs are of the tension-loaded type in which the fluid pressure stretches the disc material as the fluid pressure increases.

The continuing effort to raise the operating ratio of rupture discs led to the development of reverse-buckling discs. This type of disc is domed against the fluid pressure so that the fluid pressure introduces a compression load on the convex side of the disc.

Terminology

For the purpose of this book, the following terms are defined below:

Rupture disc device. A non-reclosing pressure relief device, consisting of rupture disc and holder, in which the rupture disc is designed to burst at a predetermined differential pressure across the disc.

Rupture disc. The pressure-containing, pressure and temperature sensitive element of a rupture disc device.

Forward-domed rupture disc. A rupture disc that is domed in the direction of the fluid pressure and designed to burst due to tensile forces.

Reverse-buckling disc. A safety disc that is domed against the direction of the fluid pressure and designed to buckle due to compression forces prior to bursting or to being expelled from the holder.

Holder. The component of the rupture disc device that holds the rupture disc around its circumference and consisting of the inlet and outlet holder parts.

Vent panel. A low-pressure venting device designed to vent the near instantaneous volumetric and pressure changes resulting from dust, gas, or vapor deflagrations.

Vacuum support. A device that supports the rupture disc against collapse due to vacuum pressure.

Back-pressure support. A device that supports the rupture disc against collapse due to superimposed back pressure.

Heat shield. A device that shields the rupture disc from the heat source in a manner that does not interfere with the rupture disc operation.

Specified temperature of rupture disc. The temperature at which the disc is rated and marked.

Burst pressure. The differential pressure across the rupture disc at which the rupture disc bursts at the specified temperature.

Marked or rated burst pressure.[1] The burst pressure at the specified temperature that is marked on the disc tag by the manufacturer. The marked burst pressure may be any pressure within the manufacturing range, unless otherwise specified by the customer.

Maximum marked burst pressure. The marked burst pressure at the top end of the manufacturing range.

Minimum marked burst pressure. The marked burst pressure at the bottom end of the manufacturing range.

Burst tolerance:[2] The maximum variation in burst pressure from the marked burst pressure.

Manufacturing range.[3] A range of pressures within which the average burst pressure of test discs must fall to be acceptable for a particular application, as agreed between the customer and manufacturer.

Performance tolerance.[4] A range of burst pressures comprising manufacturing range and burst tolerance at the specified temperature.

Operating ratio. The ratio of the maximum operating pressure to a minimum burst pressure

Damage ratio. The ratio of the burst pressure of the damaged rupture disc to the burst pressure of the undamaged rupture disc.

Reversal ratio. The ratio of the burst pressure of the reversed installed rupture disc to the burst pressure of the correctly installed rupture disc.

Lot. A quantity of rupture discs made as a single group of the same type, size, and limits of burst pressure and coincident temperature that is manufactured from material of the same identity and properties.

Deflagration. Burning that takes place at a flame speed below the velocity of sound in the medium.

Detonation. Propagation of a combustion zone at a velocity that is greater than the speed of sound in the unreacted medium.

---

[1] ASME Code (1992) Sect. VIII, Div. 1, UG 127 (a)(1)(b).
[2] ASME Code (1992) Sect. VIII, Div. 1, UG 127 (a)(1)(a).
[3] ASME Code (1992) Sect. VIII, Div. 1, UG 127 (a)(1)(a).
[4] ISO Standard 6718 and European Standards EN 1286-2 and 6.

Explosion. The bursting or rupture of an enclosure or a container due to the development of internal pressure from a deflagration.

Application of Rupture Discs

Rupture discs do not reclose after bursting. The decision to install rupture discs may therefore have important economical consequences. However, there are many applications where rupture discs are likely to perform better than pressure relief valves. These include:

• Under conditions of uncontrolled reaction or rapid overpressurization in which the inertia of a pressure relief valve would inhibit the required rapid relief of excess pressure.
• When even minute leakage of the fluid to the atmosphere cannot be tolerated at normal operating conditions.
• When the fluid is extremely viscous.
• When the fluid would tend to deposit solids on the underside of the pressure relief valve disc that would render the valve inoperable.
• When low temperature would cause pressure relief valves to seize.

Rupture discs may serve special requirements by mounting two discs in series, or in parallel, or in series with pressure relief valves

1. **Two discs in series:**
    When the process fluid may corrode the first disc, causing the discs to leak, the second disc prevents the leaking fluid from escaping to the surroundings. However, should the first disc burst prematurely, the second disc is likely to burst also.
    They also serve as a quick-opening device. By appropriately choosing the burst pressures and pressurizing the space between the discs, dumping the pressure between the discs will cause the discs to burst within milliseconds.
2. **Rupture disc in parallel with a pressure relief valve:**
    Rupture discs may be used in parallel with pressure relief valves to serve as a secondary pressure relief device that is set to protect a pressure system against overpressure excursions.
3. **Rupture disc in series upstream of a pressure relief valve:**
    Rupture discs in series are used

    • To prevent corrosive fluid from leaking into the valve. This may allow the valve to be made of standard construction materials.

- To prevent leakage past the disc of the pressure relief valve to the atmosphere or vent system.
- To prevent deposits from forming around the valve seat that would impair the operation of the pressure relief valve.
- To reduce the cost of maintaining the pressure relief valve.

4. **Rupture disc downstream of the pressure relief valve:**
   These discs are used to prevent corrosive fluids in the vent system entering and corroding the valve.

5. **Rupture discs upstream and downstream of the pressure relief valve:**
   These discs are used to combine the advantages of upstream and downstream installation of rupture discs.

Limitations of Rupture Discs in Liquid Systems

When ductile rupture discs burst in gas service, the expanding gas forces the disc open in milliseconds.

When used in liquid service, ductile rupture discs will burst in this manner only if there is a large enough gas pocket between the liquid and the rupture disc. If rapid full opening in liquid service is required, the minimum gas volume to be maintained upstream of the rupture disc is commonly recommended to be equivalent to at least 10 diameters of pipe to which the rupture disc is connected.

If the system is totally full of liquid and excess pressure is due to thermal expansion, the pressure will initially only deform the rupture disc. The resultant volume increase of the pressure system may be sufficient to initially prevent any further pressure rise. If the system pressure continues to rise, forward-acting rupture discs become finally so highly stressed that they fail at their rated burst pressure.

In the case of reverse-buckling discs, however, only a limited number of types are capable of bursting in liquid-full systems at the rated pressure. When planning to employ reverse buckling discs in liquid full systems, the manufacturer should be consulted on the selection.

Graphite rupture discs, being brittle, give instantaneously full opening upon bursting, irrespective of the type of service.

Independent of the type of rupture disc, the maintenance of a gas pocket in liquid service is advantageous for other reasons. The gas pocket minimizes the pressure rise due to volume change of the liquid and dampens peak impulse loads in pulsating service, resulting in a reduction in the frequency of disc failure.

Construction Materials of Rupture Discs

Table 6-1, Ductile Construction Materials, shows a range of ductile construction materials used for rupture discs.

Table 6-1
Ductile Construction Materials

Commonly used materials		Less commonly used materials	
Stainless steel	Aluminum	Tantalum	Platinum
Inconel	Nickel	Gold	Silver
Monel	Hastelloy B and C	Titanium	

Titanium and Hastelloy should be selected only if there is no substitute material available, as both materials tend to fail prematurely due to brittleness.

Rupture discs of ductile material may also be provided with protective coatings or linings, as shown in Table 6-2.

Table 6-2
Protective Linings and Coatings

Common protective coatings			Common protective linings		
FEP	TFE	PFA	FEP	TFE	PFA
Epoxy	Vinyl		Lead	Polyethylene	

Table 6-3 gives maximum temperatures for ductile materials, coatings, and linings.

Table 6-3
Recommended Maximum Temperatures

Ductile materials			Coatings and linings		
Aluminum	125°C	260°F	Lead	120°C	250°F
Silver	125°C	260°F	Polyvinylchloride	80°C	180°F
Nickel	425°C	800°F	FEP	215°C	400°F
Monel	425°C	800°F	TFE or PFA	260°C	500°F
Inconel	535°C	1000°F			
Stainless steel	480°C	900°F			

Data on ductile materials, coatings, linings and recommended maximum temperatures courtesy of Continental Disc Corporation.

Rupture discs of brittle material are made almost entirely of graphite, although cast iron and porcelain have been used or tried.

The graphite commonly used for rupture discs is made from low ash petroleum cokes, calcined at high temperatures. It is then mixed with pitch, formed into blocks, and then heat treated. The result is porous, brittle material that requires sealing for use in rupture discs. This is commonly done by impregnating the graphite under vacuum with either phenolic or furane resins.

Less frequently used is pure graphite. This is exfoliated graphite, originally in powder form. When suitably compressed, the graphite forms into an impervious flexible sheet. Restricted to some instances, however, pure graphite will absorb some liquid. This problem can be overcome by applying a suitable coating to the process side only. The maximum permissible operating temperature of the disc is limited in this case by the temperature resistance of the coating.

Temperature and Burst Pressure Relationships

Temperature influences the strength of the disc materials so that there is a relationship between temperature and burst pressure. The relationship varies between rupture discs of identical material but different construction.

The temperature and burst pressure relationships shown in Figure 6-1 apply to solid-metal forward-domed rupture discs as made by one manufacturer. Specific relationships are derived by the manufacturer with each lot of material.

Reverse-buckling discs are considerably less affected by temperature than forward-domed rupture discs.

Heat Shields

Heat shields are designed to shield the process side of the rupture disc from heat radiation, or heat radiation and convection. They must be installed in a manner that does not interfere with the rupture disc operation.

A heat shield may consist of overlapping stainless-steel flats that permit pressure to build up on both sides of the flats. On rupture of the disc, flow folds the flats open. The heat shield may be supplemented by a spool piece serving as a heat sink between the rupture disc and the heat shield. Further heat shielding may be provided by mounting a second heat shield to the other end of the spool piece.

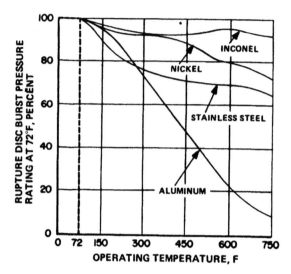

Figure 6-1. Temperature/Burst Pressure Relationship of Solid-Metal Forward-Domed Rupture Discs Made of a Variety of Materials. (*Courtesy of Continental Disc Corporation*)

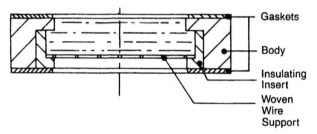

Figure 6-2. Heat Shield Consisting of Holder Filled with Loose Wool of Amorphous Silica Filaments, Intended for the Heat Shielding of Graphite Rupture Discs. (*Courtesy of IMI Marston*)

The heat shield shown in Figure 6-2 is intended for the heat shielding of graphite rupture discs. Upon bursting of the rupture disc, the heat shield will fragment and be discharged. For this reason, the heat shield cannot be employed in conjunction with a vacuum or back-pressure support that could be blocked by the fragments of the heat shield. This particular type of heat shield is suitable for dry gases only as moisture absorption changes the density of the wool filling that could lead to the collapse of the filling.

Rupture Disc Application Parameters

Rated burst pressure. The rated burst pressure, also referred to as the marked burst pressure, is the average burst pressure that has been established by bursting a minimum of two rupture discs.[1] The burst pressure thus established shall not exceed the maximum allowable burst pressure of the vessel level as defined by the code.[2]

Burst tolerance.[3] The ASME Code allows a tolerance around the rated burst pressure within which the rupture disc is permitted to burst. For burst pressures up to 2.76 barg (40 psig), the burst tolerance shall not exceed plus/minus 0.14 barg (2 psig). For burst pressures above 2.76 barg (40 psig), the burst tolerance shall not exceed plus/minus 5%. For certain types of rupture discs, manufacturers are able to offer reduced burst tolerances down to plus/minus 2%.

The burst tolerance allows the rupture disc to burst above the applicable maximum allowable burst pressure level by the amount of the burst tolerance.

Manufacturing range. The manufacturing range is an allowable range of pressures around a specified burst pressure within which the rupture disc can be rated. The pressure range must be agreed upon between user and manufacturer. The purpose of the agreement is to permit the economical production of some types of rupture discs.

The types of rupture discs mainly involved are forward-domed rupture discs, which by their design fail in tension. Because the tensile strength of the metals used in the manufacture of rupture discs is fairly high, the discs must be made of relatively thin foils. Thin foils of uniform thickness and tensile strength, however, are difficult to produce and vary between heats of material.

Manufacturers must therefore select from different foils until the desired burst pressure has been achieved. To keep manufacturing costs within acceptable limits, only a limited number of finding tests can be carried out.

The manufacturing range may be expressed as a minus or plus/minus percentage around the specified burst pressure or in pressure units. These are two examples.

---

[1] ASME Code (1992) Sect. VIII, Div. 1, UG 127(a)(1)(a).
[2] ASME Code (1992) Sect. VIII, Div. 1, UG 134.
[3] ASME Code (1992) Sect. VIII, Div. 1, UG 127(a)(1).

If the specified burst pressure is 10 barg (145 psig) and the manufacturing range is stated to be minus 11%, a disc that is rated anywhere between 10 barg (145 psig) and 8.9 barg (129 psig) meets the disc specification.

If the specified burst pressure is 10 barg (145 psig), as before, but the manufacturing range is stated to be plus 7%/minus 4%, a disc that is rated anywhere between 10.7 barg (155 psig) and 9.6 barg (139 psig) meets the disc specification. The maximum rated burst pressure must not exceed the maximum allowable burst pressure except where permitted by the code.

Manufacturers are able to offer reduced manufacturing ranges and, in the case of reverse-buckling discs, also zero manufacturing range.

Operating ratio. This is the ratio between the maximum operating pressure and the minimum burst pressure. The operating ratio is designed to ensure a satisfactory service life of the rupture disc. The values recommended by manufacturers range between 70% or less and up to 90%, depending on type of rupture disc and operating temperature. Non-steady operating conditions, such as cycling and pulsating pressures and fluctuating operating temperatures, may vary these values.

METAL RUPTURE DISCS

There are two major types of rupture discs made of ductile metal:

* forward-acting types, being tension loaded
* reverse-acting types, being compression loaded

Forward-domed and flat rupture discs are the tension-loaded types, while the reverse-buckling disc is of the reverse-loaded type. The following describes a cross-section of these discs as offered by the industry.

Tension-Loaded Types

Solid forward-domed rupture discs. Solid forward-domed rupture discs are formed from flat discs by applying a fluid pressure to the underside of the disc of normally above 70% of the burst pressure. This method of manufacture gives the rupture disc a hemispherical shape, as shown in Figure 6-3. When operating pressure grows beyond the predoming pressure, the dome starts to grow. As the operating pressure approaches 95% of the burst pressure, localized thinning in the region of the dome center

Figure 6-3. Solid Forward-Domed Rupture Disc Before and After Bursting in Gas Service. (*Courtesy of Continental Disc Corporation.*)

occurs that leads to rupture of the disc. This failure is accompanied by some fragmentation of the disc.

To guard the disc against further plastic deformation during service, normal operating pressure is commonly restricted to 70% of the rated burst pressure or less, depending on operating conditions.

Because the tensile strength of the construction material used for the manufacture of the discs is fairly high, solid forward-domed rupture discs for low pressures must be made of relatively thin foils. Periods of vacuum or superimposed back pressure will therefore tend to cause the disc to partially or fully collapse. When these conditions exist, the disc must be provided with vacuum supports such as those shown in Figure 6-4, or in special cases, such as superimposed back pressure, with supplementary permanent supports as shown in Figure 6-5. These supports must fit closely the concave side of the disc to prevent alternating collapsing and stretching of the disc. The deformation of disc manifests itself in a wrinkle pattern identified as turtle backing, as shown in Figure 6-6, resulting in a poor service life.

Figure 6-3 shows one of the discs after bursting in gas service. In full-liquid service, the rupture disc may burst initially in a pattern as shown in Figure 6-7 and open further with rising overpressure.

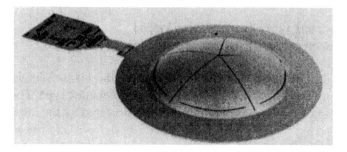

Figure 6-4. Vacuum Support for Forward-Domed Rupture Discs, Full Opening Upon Bursting of the Disc. (*Courtesy of IMI Marston Ltd.*)

Figure 6-5. Permanent Back Pressure Support to Supplement Vacuum Support. (*Courtesy of IMI Palmer Ltd.*)

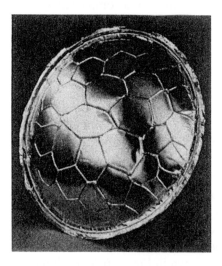

Figure 6-6. Wrinkle Pattern Identified as Turtle Backing of Solid Forward-Domed Rupture Discs Due to Alternate Reverse Flexing and Stretch Against Inadequate Support. (*Courtesy of Continental Disc Corporation*)

Figure 6-8 shows a solid forward-domed rupture disc for low-pressure applications in which the thickness of the burst element is at the near minimum. The seatings of the burst element are supported on both sides by protective rings that carry on the concave side an integral vacuum support

Figure 6-7. Solid Forward-Domed Rupture Disc Showing Rupture Pattern after Bursting in Liquid-Full System. (*Courtesy of BS&B.*)

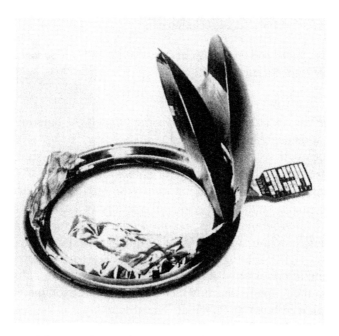

Figure 6-8. Solid Forward-Domed Rupture Disc for Low Fluid Pressures, Provided with Vacuum Support and Protection Cap. (*Courtesy of Rembe GmbH.*)

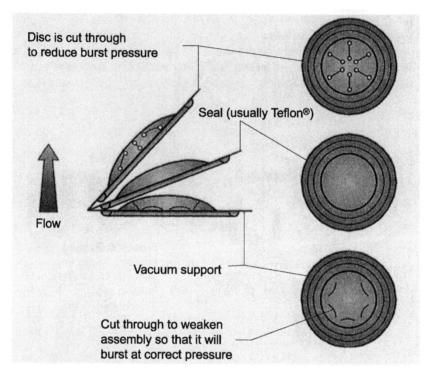

Figure 6-9. Slotted and Lined Forward-Domed Rupture Disc. (*Courtesy of Anderson, Greenwood & Company.*)

and on the convex side an integral cap that protects the bursting disc from inadvertent damage during handling.

Advantages of solid forward-domed rupture discs are simple design, more cost-effective, and suitability for liquids and gases.

Disadvantages are the operating ratio is limited to 70% or less; they are not normally suited for periods of vacuum or back pressure unless provided with vacuum or back-pressure support; they may not be suitable for pulsating pressure; and the disc may fragment upon failure.

Slotted and lined forward-domed rupture discs. This is a multi-layered forward-domed rupture disc in which the dome of the top member is slotted with pierced holes at each end. The second layer is the seal member, commonly made of fluorocarbon or an exotic metal. A vacuum support that may be required is the third component. Figure 6-9 illustrates the three layers of the disc.

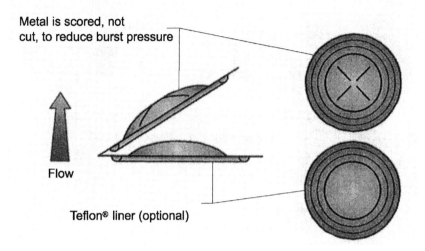

Metal is scored, not
cut, to reduce burst pressure

Flow

Teflon® liner (optional)

Figure 6-10. Cross-Scored Forward-Acting Rupture Disc. (*Courtesy of Anderson, Greenwood & Company.*)

For a given thickness and strength of the material, the burst pressure is controlled by a combination of slits and tabs. By this construction, the rupture disc for low burst pressures can be manufactured from thicker materials that permit the operating ratio to be raised to 80%.

The rupture discs are generally suitable in the lower pressure regions only. They may be used in gas and liquid service and permit pulsating pressure service. The discs are also non-fragmenting when used in conjunction with a fluorocarbon seal member.

Scored forward-domed rupture discs. These are solid forward-domed rupture discs that are cross scored on the convex side of the dome, as shown in Figure 6-10. Scoring allows the disc to be made of thicker material that allows the operating ratio to be raised to 85%. The discs may be used in either gas or liquid-full service and offer a good service life in cycling service. The score lines provide a predictable opening pattern so that the disc can be manufactured to be non-fragmenting. Within the lower burst pressure range, however, the rupture disc must be supported against full vacuum.

A special field of application of scored forward-acting rupture discs is in polymer service. The problem of polymerization is minimized by avoiding a crevice between the rupture disc and inlet holder component in which growth of polymer could start. A rupture disc device that meets

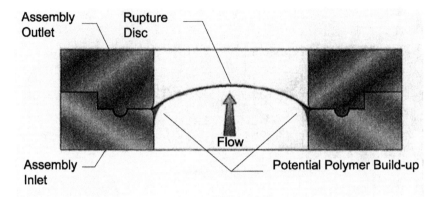

Assembly Outlet

Rupture Disc

Flow

Assembly Inlet

Potential Polymer Build-up

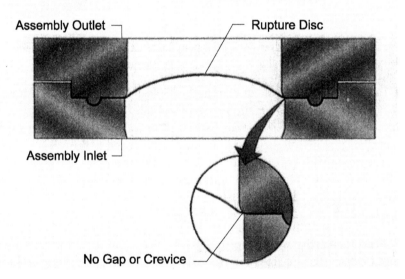

Assembly Outlet

Rupture Disc

Assembly Inlet

No Gap or Crevice

Figure 6-11. Rupture Disc Device Consisting of Cross-Scored Forward-Domed Rupture Disc and Holder Designed for Polymerization Service. (*Courtesy of Anderson, Greenwood & Company.*)

this requirement is shown in Figure 6-11. By implication, the rupture disc cannot be used in polymerization service when being fitted with a vacuum support.

Figure 6-12 and Figure 6-13 show two types of slotted and lined flat rupture discs that may be used for either one-way or two-way flow.

Figure 6-12. Slotted and Lined Flat Rupture Disc Used as an Environmental Seal For Transport and Storage Tanks and Downstream of Pressure Relief Valves. (*Courtesy of Continental Disc Corporation.*)

The rupture disc shown in Figure 6-12 is a low-cost pressure relief device typically used as an environmental seal for transport and storage tanks and downstream of pressure relief valves. The disc does not require separate holders but can be mounted directly between class 150 flanges. Variations are designed for vacuum relief only, or for pressure in one direction and for maximum double that pressure in the other direction. Highest operating ratio is restricted to 50%.

The rupture disc shown in Figure 6-13 is designed for overpressure relief of low-pressure systems. The discs withstand full vacuum and may be used in gas and liquid-full systems. Operating ratio of the disc is as high as 80%.

Compression-Loaded Types

In reverse buckling discs, the fluid load acts on the convex side of the disc. This loading puts the disc in compression.

The material property that determines the buckling pressure is the Young's modulus. This property is much more constant and reproducible, and also less affected by temperature than the ultimate tensile strength of metals used in the construction of rupture discs. In addition, buckling occurs at substantially lower stress level than rupture under tensile stress. Reverse buckling discs must therefore be made of a considerably thicker material than forward-domed rupture discs. Consequently, they are much

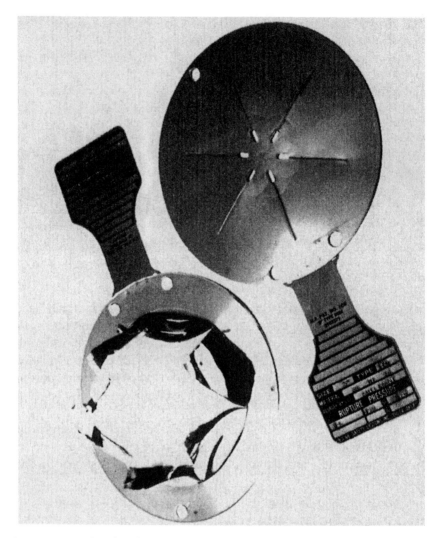

Figure 6-13. Slotted and Lined Flat Rupture Disc for Overpressure Protection of Low-Pressure Systems. (*Courtesy of BS&B.*)

easier to produce to close tolerances over a wide temperature range than rupture discs that fail in tension.

The buckling pressure is determined not only by the properties of the material but also by the shape of the dome. When the disc is exposed to rising temperature, the strength of the material falls while the dome expands and gains strength. This gain in strength partially compensates for loss in material strength due to rising metal temperature. Buckling discs are, therefore, less sensitive to temperature changes than forward domed rupture discs.

Because reverse-buckling discs function at low stress levels, there is no permanent deformation until the disc starts to buckle. This buckling process proceeds exceedingly fast. By itself, the disc does not burst open on reversal. This is achieved either by a cutting device against which the disc must be slammed, or by scoring the disc, or by expelling the disc from the holder.

Outstanding advantages of reverse-buckling discs are low burst pressure capabilities; operating ratio up to 90% and higher; on request, zero manufacturing range and reduced burst tolerance, excellent for cyclic and pulsating pressures; extended service life due to being less affected by fatigue than forward-domed rupture discs.

The following shows a cross section of numerous variants of reverse-buckling discs that have been developed.

Reverse buckling disc with knife blades. Figure 6-14 shows a reverse buckling disc in combination with knife blades that are designed to cut the disc open upon reversal. For this to happen, the disc must strike the knife blades with high energy. The disc may therefore be used in gas service only and in liquid service if there exists a substantial gas volume between the liquid and the disc. In totally full-liquid systems, reversal speed will be too slow to cut the disc. In this case, the disc comes initially to rest on the knife blades before being cut open after a substantial pressure rise. In the past, this situation has led to a number of recorded pressure vessel ruptures.

It is essential that the edges of the knife blades are kept sharp. The cutting edges must therefore be checked on a regular basis and must be resharpened if necessary. Care must be taken not to change the blade location or configuration. In most cases, the manufacturer should perform repair or replacement.

The advantages of this type of disc are that it can be designed for low burst pressures, it does not require vacuum support, it is excellent for cyclic or pulsating pressures, it is non-fragmenting, and it may be offered for 90% operating ratio, zero manufacturing range, and plus/minus 2% burst tolerance.

Its disadvantages are the knife blades must be kept sharp, and it is not suitable for liquid-full systems.

Reverse-buckling disc with teeth ring. The reverse-buckling disc shown in Figure 6-15 is provided with a teeth ring that pierces and cuts the rupture disc on buckling. The disc offers advantages and disadvantages similar to

Figure 6-14. Exploded View of Reverse-Buckling Disc with Knife Blades. (*Courtesy of Continental Disc Corporation.*)

the reverse-buckling disc with knife blades, but is designed for considerably lower burst pressures.

Reverse-buckling disc cross-scored. Figure 6-16 shows a rupture disc that is cross-scored on the concave side. Upon buckling, the disc will break open in pie-shaped sections along the score lines, with the base firmly held by the holder.

The cross-scored rupture disc has the advantage over the reverse-buckling disc shown in Figure 6-14 by functioning without the assistance of knife blades. The advantages and disadvantages of both types of reverse-buckling discs are otherwise identical.

Reverse-buckling disc with partial circumferential score line. Figure 6-17 through Figure 6-19 show three types of reverse-buckling discs with a partial circumferential score line around the rim of the disc. When buckling occurs, the discs shear open along the score line and fold around a pivot.

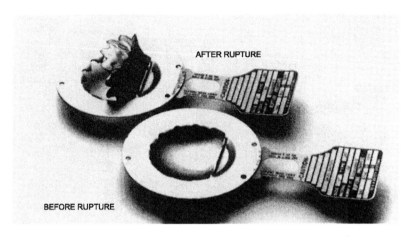

Figure 6-15. Reverse-Buckling Disc With Teeth Ring. (*Courtesy of BS&B.*)

Figure 6-16. Reverse-Buckling Disc, Cross-Scored and Annealed, Mounted in Prelorque-Type Holder, Before and After Rupture. (*Courtesy of BS&B.*)

Figure 6-17. Reverse-Buckling Disc with Partial Circumferential Score Line. (*Courtesy of Continental Disc Corporation.*)

Figure 6-18. Reverse-Buckling Disc With Partial Circumferential Score Line, Low Pressure Series. (*Courtesy of Fike Metal Products.*)

In the case of the rupture disc shown in Figure 6-19, the score is perforated to permit the achievement of an extremely low burst pressure. The perforations are sealed with an O-ring that becomes energized as soon the fluid pressure is applied.

These rupture discs offer all the advantages of cross-scored rupture discs. In addition, the discs may be employed in liquid-full systems.

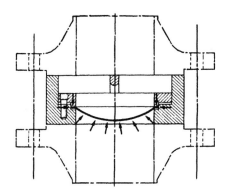

Figure 6-19. Reverse-Buckling Disc Partial Circumferential Perforated Score, Designed for Lowest Pressures. (*Courtesy of BS&B.*)

Figure 6-20. Reverse-Buckling of Slip-Away Design That Slips Out of the Holder Parts on Buckling Without Rupture. (*Courtesy of IMI Marston Ltd.*)

Reverse-buckling disc of slip-away design. Reverse-buckling discs of the slip-away design, as shown in Figure 6-20, function by being expelled from the holder upon buckling. The relatively narrow flat seatings of the disc are mounted in a recess of the inlet holding part and sealed with an O-ring or a flat gasket. No special torque settings are required. To prevent the disc from traveling along the vent line, the holder can be provided with an integral or separate arresting device. This allows the disc to be mounted upstream of the pressure relief valve.

Advantages of this design are the operating ratio can be up to 95%; the reversal ratio in general is less than 1.0 with a maximum of 1.1 for smaller sizes; and the size range is from DN 25 (NPS 1) to DN 1200 (NPS 48).

The disadvantages are it is not suitable for use in liquid-full systems except in the larger sizes, and it may require vacuum support.

Reverse-buckling disc slotted and lined with buckling bars. The rupture disc shown in Figure 6-21 consists of a slotted component that is the actual pressure sensitive element, and a seal element of either plastic or metal. In combination with a metal seal, the disc supports full vacuum.

The partial circumferential slot is teeth-shaped, while the tabs represent buckling bars that control the buckling process. On reaching the buckling

Figure 6-21. Slotted and Lined Reverse-Buckling Disc with Buckling Bars as Burst Control Element. (*Courtesy of Rembe GmbH.*)

pressure, the bars buckle and break off and allow the disc to hinge open. During this process, the teeth ring cuts open the seal member.

The disc is made in sizes DN 25 (NPS 1) through DN 600 (NPS 24). The burst pressure range covers the lowest burst pressures up to 120 barg (1800 psig) at 22°C (72°F) operating temperature. Maximum operating temperature is 550°C (1000°F).

The advantages of this disc are zero manufacturing range, burst tolerance is down to plus/minus 2%, operating ratio is as high as 95%, and it may be employed in liquid-full systems (consult manufacturer).

Two-way rupture disc. In the case of domed two-way rupture discs, such as shown in Figure 6-22 and Figure 6-23, flow in one direction is forward acting and in the other direction reverse acting.

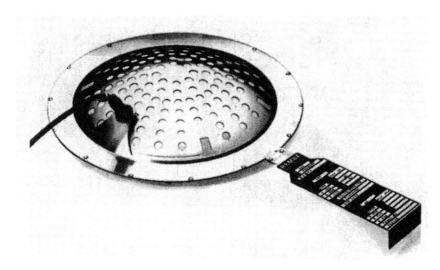

Figure 6-22. Two-Way Rupture Disc Incorporating Buckling Bars. (*Courtesy of Rembe GmbH.*)

The rupture disc shown in Figure 6-22 consists of three components. The upper perforated component is a forward-domed rupture disc that is provided with a partial circumferential score line and protects the pressure system against overpressure. The second element is a PTFE seal member that fails under overpressure and subnormal pressure conditions. The third element is represented by a reverse-buckling bar disc that fails on subnormal pressure. On rupture of the upper component, the buckling bars will fail in tension and allow the disc to open in unity with the upper disc.

The two-way rupture disc shown in Figure 6-23 consists basically of a perforated forward-acting rupture disc that is sealed by a Teflon® reverse-acting disc, followed by a girdle supporting the disc and knife blades. Under conditions of subnormal pressure, the Teflon® disc will be forced against the knife blades and be cut. Under conditions of excessive overpressure, the perforated disc and the supporting Teflon® disc will be put in tension and rupture.

Graphite Rupture Discs

Graphite is a valuable material from which low-pressure rupture discs can be produced.

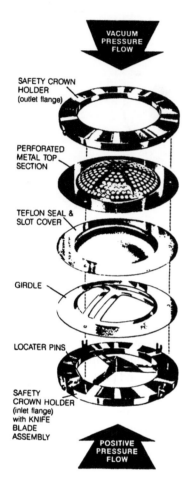

SAFETY CROWN HOLDER (outlet flange)

PERFORATED METAL TOP SECTION

TEFLON SEAL & SLOT COVER

GIRDLE

LOCATER PINS

SAFETY CROWN HOLDER (inlet flange) with KNIFE BLADE ASSEMBLY

VACUUM PRESSURE FLOW

POSITIVE PRESSURE FLOW

Figure 6-23. Two-Way Rupture Disc Incorporating Reverse-Acting Teflon® Disc with Knife Blades. (*Courtesy of Continental Disc Corporation.*)

The graphite commonly used for the manufacture of rupture discs is the resin-impregnated grade. The material is very brittle and ruptures almost without deformation. Its structure is very homogeneous and the strength of the material is low compared with the strength of metals. Graphite rupture discs are therefore much thicker than metal rupture discs. This property permits graphite rupture discs to be made to small burst pressure tolerances.

Pure graphite is a flexible material that is used for the manufacture of reverse-buckling discs. Because pure graphite is free of resin impregnation, pure graphite rupture discs may be exposed to higher operating temperatures.

Graphite rupture discs may be used in both gas and liquid-full services.

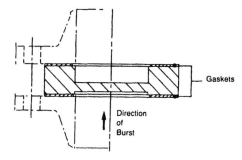

Figure 6-24. Monoblock Type Rupture Disc Made of Brittle Graphite, Non-Armored Construction. (*Courtesy of IMI Marston Ltd.*)

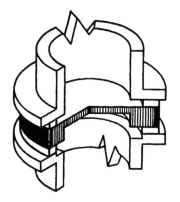

Figure 6-25. Monoblock-Type Rupture Disc Made of Brittle Graphite, Armored by a Steel Ring Bonded to the Disc Circumference. (*Courtesy of Continental Disc Corporation.*)

Monoblock-type graphite rupture discs. Rupture discs made of resin-impregnated graphite are commonly produced in monoblock form, as shown in Figure 6-24 through Figure 6-26. These are one-piece devices that combine a flat bursting membrane with the mounting flange.

The rupture disc shown in Figure 6-24 is of non-armored construction made in sizes DN 25 (NPS 1) through DN 600 (NPS 24). Operating ratio is 90%. The discs are suitable for operating temperatures ranging from minus 70°C (minus 94°F) to 180°C (356°F). In conjunction with a heat shield shown in Figure 6-2, the operating temperature may be raised to 500°C (930°F). The rupture disc, however, requires a controlled torque loading of the flange bolts.

The rupture disc shown in Figure 6-25 is armored by a steel ring bonded to the disc circumference to prevent unequal piping stresses from reaching the pressure membrane of the disc.

The rupture disc shown in Figure 6-26 is a two-way version of graphite rupture discs. The discs are custom-produced in sizes DN 25 (NPS 1)

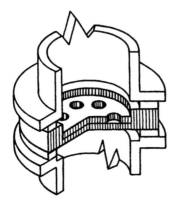

Figure 6-26. Rupture Disc of Brittle Graphite, Two-Way Type. (*Courtesy of Continental Disc Corporation.*)

through DN 600 (NPS24). The maximum burst pressure in either direction is 10 bar (150 psig). The minimum pressure differential between the two burst pressure ratings must be 0.7 barg (10 psig), depending upon diameter and rating of the rupture disc.

Rupture disc device with replaceable graphite bursting membrane. The rupture disc device shown in Figure 6-27 represents a design in which the graphite burst membrane is a replaceable component. The purpose of this design is to achieve a greater economy of graphite. The reverse pressure supports shown, however, reduce the available vent area to approximately 50% of full-bore area.

The disc is mounted in a controlled depth recess. By this arrangement, the installation of the disc is non-torque sensitive.

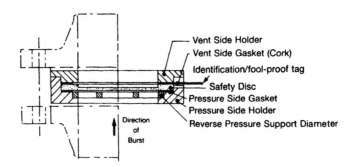

Figure 6-27. Rupture Disc Device with Replaceable Disc of Brittle Graphite, Non-Torque Sensitive Construction. (*Courtesy of IMI Marston Ltd.*)

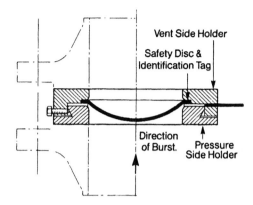

Figure 6-28. Rupture Disc Device with Reverse-Buckling Disc of Pure Graphite. (*Courtesy of IMI Marston Ltd.*)

Reverse-buckling rupture disc of pure graphite. The disc, such as shown in Figure 6-28, is made of pure graphite without the inclusion of resin. For this reason, the disc can be used for chemicals and high temperatures [typically 550°C (1000°F)] that would normally affect resin impregnated graphite rupture discs. Because the discs are thicker than the membrane of the equivalent resin-impregnated graphite rupture discs, reverse-buckling rupture discs of pure graphite can easily be handled.

Discs for burst pressures above 1.2 barg (17.4 psig) will support full vacuum. Below this pressure, an additional support may be required. For burst pressures higher than 1.0 barg (14.5 psig), the burst tolerance is plus/minus 5%.

Rupture Disc Holders

The original disc holders were intended for metal rupture discs that failed in tension. They consisted of two flat-faced flanges between which the discs were clamped.

This design encountered two problems. First, the disc would tend to shear on the sharp edge of the top flange, resulting in unpredictable failure. Second, as pressure increased, the disc would tend to slip between the flat faces, resulting in leakage and unpredictable failure.

The first problem was overcome by radiusing the inside corner of the top flange. To overcome the second problem, designers changed the flat seat to a 30° conical seat that wedges the disc between two faces, as shown in Figure 6-29.

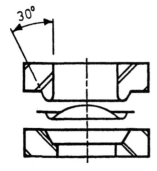

Figure 6-29. Exploded View of Rupture Disc Holder for Solid-Metal Forward-Domed Rupture Discs. (*Courtesy of Continental Disc Corporation.*)

This configuration still has a debit side. Overtightening of the conical seat can cause thin foils to thin out or rupture, leading to early failure of the disc. For this reason, the seatings of fragile rupture discs are often protected by stiffening rings, not only to prevent damage to the disc during handling, but also to minimize damage to the 30° angle disc faces on tightening.

Overtightening can also deform the conical lip of the outlet holder part. This failure has been overcome for high pressure applications by providing a heavy conical lip. For best performance of the rupture disc, the torque recommendations of the manufacturer must be observed.

For most rupture disc manufacturers in the U.S., the 30° conical seat configuration is standard. A deviation from this practice is the flat seat configuration with tongue and groove to accommodate the flanges of rupture discs, such as those shown in Figure 6-3 and others. In many parts of Europe and the Far East, the flat seat has been retained for the lower pressure applications because of the advantage of not thinning the very thin foils on the conical seat.

Because reverse buckling discs are loaded in compression, they do not present the problem of slippage between the holder parts so that the seating faces can be flat regardless of pressure. It is important, however, that the supporting edges of the holder do not deform the dome or the edges of the dome. Any such deformation would affect the bursting pressure.

Rupture disc holders in flanged pipelines are commonly designed for mounting between two flanges. This allows the rupture disc to be inserted in the workshop under clean conditions. The holder parts, however, are offered also with weld-neck connections. The user should be aware that, with this construction, the insertion of the rupture disc has to be carried out on site often under unfavorable conditions.

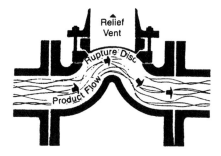

Figure 6-30. Clean-Sweep Design of Rupture Disc Device, Incorporating Forward-Domed Rupture Disc. (*Courtesy of Continental Disc Corporation.*)

Clean-Sweep Assembly

The clean-sweep assembly, as shown in Figure 6-30, also known as viscous Tee, is intended for applications in which viscous fluid could cause product build-up and plugging of the flow passage. The sweeping flow passage reduces this problem.

Quick-Change Housings

To permit rapid replacement of burst rupture discs, a number of quick-change devices have been developed such as shown in Figure 6-31 and Figure 6-32. One of these devices requires change-over to be carried out by hand. The second device is designed for change-over by means of a pneumatic actuator.

Sealing of the positioned rupture disc is affected by O-rings mounted in the change-over housing. Once the system pressure is restored, the O-rings provide a powerful interlock between the change-over housing and the rupture disc holders.

Accessories

The following lists a number of accessories that are being supplied by manufacturers in connection with rupture disc holders.

Jackscrews. These are used as a means for separating the flange carrying the disc holder from the flange of a separately supported discharge pipe. The jackscrews must be properly spaced to allow the disc holder to be removed or reinstalled.

Figure 6-31. Quick-Change Housing. (*Courtesy of Continental Disc Corporation.*)

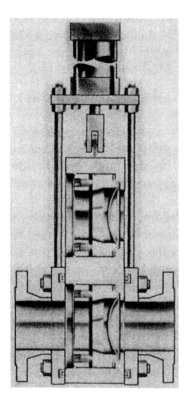

Figure 6-32. Quick-Change Housing, Pneumatically Operated Sliding Gate Construction. (*Courtesy of Continental Disc Corporation.*)

Eyebolts. These are used primarily to hoist heavy rupture disc holders into position.

Baffle plates. These are primarily used in connection with free-venting rupture disc assemblies in which deflection of the discharging fluid would protect personnel. The deflection of the discharging fluid absorbs, in addition, the discharge reactive force.

Burst indicators. These give an instantaneous indication of a rupture disc failure by actuating a visible or audible signal that may be locally or remotely located. These can be designed to be intrinsically safe within an explosively hazardous environment.

Double Disc Assemblies

Double disc assemblies, such as shown in Figure 6-33, are being used in corrosive services to warn of product leakage due to corrosion damage of the disc on the process side. In another application, double disc assemblies are used as a quick-opening device.

Product leakage warning devices. Figure 6-33 shows a double disc assembly used for warning of product leakage due to corrosion damage of the disc facing the product side. To prevent pressure build-up in the space between the rupture discs due to pinholes from corrosion in the first disc, the space between the discs must be provided with a telltale device, consisting commonly of an excess flow valve and a pressure indicator. Should the first disc fail prematurely, the second disc is likely to fail also. If this mode of failure cannot be accepted, the burst pressure of the second

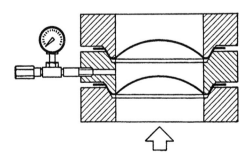

Figure 6-33. Double Disc Assembly with Product Leakage Warning Devices Consisting of Pressure Gauge and Excess Flow Valve. (*Courtesy of Sempell A.G.*)

disc must be raised. This matter should be discussed with the rupture disc manufacturer.

The ASME Code does not deal with double disc assemblies.

Quick-opening device. By lowering the rating of the rupture discs in a double disc assembly to below the operating pressure, the assembly may be used as a quick-opening device. One method of initiating quick opening is to pressurize the space between the discs until the outlet disc bursts. The operating pressure will then burst the first disc within milliseconds. The second method requires the maintenance of a pressure of approximately half the normal system pressure. On dumping this pressure, both discs open simultaneously within milliseconds.

When employing the second method, the following limitations apply:

- The pressure between the rupture discs cannot exceed the operating pressure unless the rupture disc facing the operating pressure is fitted with a back-pressure support.
- The burst pressure of either rupture disc, including burst tolerance, cannot be greater than the operating pressure.

For conditions of both rupture discs being identical, the following pressure relationships exist:

$$\frac{P_2}{P_{B2}} = \frac{P_1 - P_2}{P_{B1}} = \text{operating ratio of rupture disc}$$

$$\text{then} \quad P_2 = \frac{P_1}{2}$$

where P_1 = normal gauge operating pressure
$\quad\quad$ P_2 = gauge fluid pressure in space between rupture discs
$\quad\quad$ P_{B1} = rated burst pressure minus burst tolerance of inner disc
$\quad\quad$ P_{B2} = rated burst pressure minus burst tolerance of outer disc

Example:
Normal operating pressure $P_1 = 140$ psig
Operating ratio of rupture disc = 0.7, burst tolerance plus/minus 5%.

Fluid pressure in space between rupture discs $P_2 = \frac{P_1}{2}$ 70 psig

Then $P_{B1} = P_{B2} = \dfrac{70}{0.7} = 100$ psig minimum burst pressure
$\quad\quad\quad\quad\quad\quad\quad\quad\quad = 105$ psig rated burst pressure.

Rupture Discs as Sole Relieving Devices

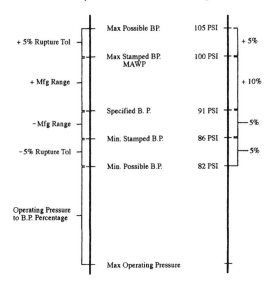

Figure 6-34. Pressure Level Relationships of Installed Rupture Disc (Typical Only), Taken from Fike Technical Bulletin TB 8100-2. (*Courtesy of Fike Metal Products.*)

Selecting Rupture Discs

The selection of rupture discs is interdependent on the pressure differential between the operating pressure and the maximum allowable working pressure of the pressure system. Manufacturing range, burst tolerance, and recommended operating ratio for the selected rupture disc must be accommodated within this pressure differential.

Figure 6-34 shows the pressure level relationships between the maximum allowable working pressure and the operating pressure resulting from the installation of a rupture disc as the sole pressure relief device. Manufacturing range in this case is plus 10%/minus 5% and the burst tolerance plus/minus 5%.

For an installation of this type, the ASME Code[1] requires the rated burst pressure not to exceed the maximum allowable working pressure but permits the plus burst tolerance to do so. Under these conditions, the minimum possible burst pressure could amount to 82% of the maximum allowable

[1] ASME Code (1992) Sect. VIII, Div. 1, UG-134.

working pressure. Allowing for an operating ratio of 70%, the operating pressure should not exceed 57% of the maximum allowable working pressure.

For burst pressures higher than 2.75 barg (40 psig), some rupture disc manufacturers deviate from the assessment of the minimum burst pressure as shown in Figure 6-34 by neglecting the minus 5% burst tolerance.

When the pressure differential is as short as 10%, as is common in pressure relief valve installations, the only available choice is a rupture disc with zero manufacturing range, an operating ratio of 90% or 95%, and possibly a reduced burst tolerance.

The selection of the rupture disc must also take account of the service temperature of the disc and possible temperature excursions, fluid state such as gas or liquid, cyclic or pulsating service, corrosiveness of environment on both sides of the disc, vacuum or back pressure, and possibly other operating conditions. Some types of rupture discs are available in a limited range of pressures and sizes only. In some cases, therefore, the manufacturer may need to be involved in the selection of rupture discs already at the early project design stage.

Rupture Disc Device in Combination with Pressure Relief Valve

Rupture discs may be used in combination with pressure relief valves to isolate valve inlet and/or outlet from corrosive fluids prior to the valve being called upon to open. This duty of the rupture disc must be carried out in a manner that assures the proper relief function of the pressure relief valve. The following discussion is restricted to installations in gas or vapor service.

Rupture disc device at inlet of pressure relief valve. The operational requirements are given below:

- The pressure relief valve shall pop open in unison with the bursting of the rupture disc.
- Leakage across the corroded rupture disc must be prevented from pressurizing the space between the rupture disc and the pressure relief valve.

In the majority of installations, the operating differential amounts to only 10% as is common for pressure relief valve installations. For this operating ratio, rupture discs with zero manufacturing range, a burst

tolerance of plus/minus 2%, and an allowable operating ratio of 90 or 95% are available.
* To ensure that the rupture disc opens fully, rupture disc and pressure relief valves should open simultaneously. But taking into account a set pressure tolerance of the pressure relief valve of plus/minus 3% and a burst tolerance of plus/minus 2%, the opening pressure of the pressure relief valve could, in the worst case, be higher than the burst pressure by 5%. To compensate for this possibility, the set pressure of the pressure relief valve may have to be 3% lower than the rated burst pressure if this does not lower the blowdown below the permissible level. Setting the valve not more than 5% higher than the disc has been reported to achieve full opening of the disc,[1] however, setting the valve lower than the rupture disc is the preferred option.

The pressure relief valve used in this combination should be of the pop-open type and may be unbalanced unless back-pressure conditions demand a balanced type.

The ASME Code[2] requires a telltale indicator assembly to be fitted in the space between the pressure relief valve and the rupture disc to monitor any leakage into this space across a corroded disc. Any pressure build-up in this space would raise the burst pressure of the disc by the amount of leakage pressure build-up.

The ASME Code permits the capacity of the combination of rupture disc and pressure relief valve to be taken at 90% of the rated capacity of the pressure relief valve or at the value of the combination factor as certified by test.[3]

Rupture disc device of outlet of pressure relief valve.[4] The purpose of this combination is to protect the valve internals against the ingress of corrosive fluids that might be present in the discharge system.

This pressure relief combination permits the entry of leakage past the valve seatings into the chamber between the seatings and the rupture disc. To prevent a pressure build-up in this chamber, the upstream side of the rupture disc must be provided with a telltale indicator assembly.

---

[1] Teledyne Farris Safety and Relief Valve Catalog No. FE-80-100 page 7.26.
[2] ASME Code (1992), Sect. VIII, Div. 1, UG-127 (3)(4).
[3] ASME Code (1992), Sect. VIII, Div. 1, UG-129 (c), UG-132.
[4] ASME Code (1992), Sect. VIII, Div. 1, UG 127 (c)(3).

When the valve is called upon to open, pressure builds up in the chamber between the seatings and the rupture disc that amounts to the burst pressure of the rupture disc. In assessing the possible burst pressure, the maximum rated burst pressure plus burst tolerance and any superimposed back pressure should be taken into account. The valve and piping components upstream of the rupture disc must be designed to withstand this pressure.

The pressure build-up thus assessed could impede the opening of a conventional pressure relief valve. For this reason, the use of a balanced pressure relief valve is suggested.

Rupture disc devices at inlet and outlet of pressure relief valve. The design may follow the design recommendation for the above rupture disc/pressure relief valve combination, choosing a balanced pop-action pressure relief valve.

Rupture disc/pressure relief valve combination for use in liquid service. Rupture disc/pressure relief valve combinations with the rupture disc mounted to the valve inlet can be used also in liquid service if a substantial gas volume between the rupture disc and the liquid level can be assured so that the pressure relief valve opens in unison with the bursting of the rupture disc.

When considering the use of rupture discs in combination with pressure relief valves, the designer should follow the applicable design code.

Reordering Rupture Discs

When reordering rupture discs, the user should always quote the lot or batch number to ensure the supply of the correct replacement disc. The user should also be aware that the requested burst pressure only needs to fall within the manufacturing range of the ordered disc to meet the manufacturer's obligations. The rated or marked burst pressure can therefore vary between orders of the same rupture disc.

User's Responsibility

The user's personnel must be fully aware of the handling, storage, installation, and maintenance requirements of rupture discs. Under no circumstances should the installation be entrusted to a person who is unaware of the installation requirements and the consequences of improper handling of rupture discs.

Each plant should maintain a system that records location and specification of installed rupture discs, the date and history of rupture disc failures, the agreed-upon replacement periods, and ordering and storage instructions. The consequences of any misunderstanding can be costly.

In general, only personnel trained in the selection, installation, and maintenance of rupture discs should be entrusted with specifying, procuring, and installing rupture discs.

Explosion Venting

Explosion Vent Panels. Many powders used in the industry as well as flammable gases form explosive mixtures when dispersed in air. On ignition, such mixtures will explode and nearly instantly cause the pressure to rise to a level that would rupture low-strength enclosures with inadequate venting.

When the risk of such explosions exists, low-strength enclosures are provided with vent panels, commonly of either rectangular or circular shape, that are sized to prevent the pressure in the enclosure from rising above a safe level. Figure 6-35 shows a typical explosion vent panel and Figure 6-36 a comparative graph of a vented and an unvented explosion. P_{stat} in the graph is the burst pressure, P_{red} is the reduced explosion pressure, P_{ult} is the ultimate sustainable pressure of the enclosure, and P_{max} is the pressure level of an unvented explosion.

Explosion vent panels in combination with flame-quenching vent system. When a dust deflagration occurs in a plant item, there is far more dust present than there is oxidant to burn the dust completely. When venting takes place, large amounts of unburned dust mixes with additional air from the surrounding atmosphere. In one deflagration venting test, a fire ball extended at least 4 m below the level of the vent and about 15 m horizontally. Personnel enveloped by such a fireball might not survive[1] (see Figure 6-37 for a random dust explosion test). The protected plant item must therefore be positioned in a manner that protects the personnel and prevents secondary explosions.

---

[1] NFPA68 (1988), Guide for Venting of Deflagrations (7-7, Flame Clouds from Deflagrations).

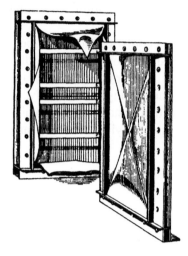

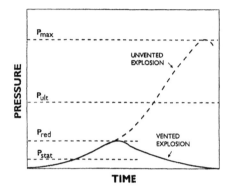

Figure 6-35. Explosion Vent Panel of Composite Construction Before and After Bursting. (*Courtesy of Continental Disc Corporation.*)

Figure 6-36. Comparative Graph of Vented and Unvented Explosion. (*Courtesy of Fike Metal Products.*)

Figure 6-37. Test Showing Ability of Flames to Propagate through Piping. (*Courtesy of Fike Metal Products.*)

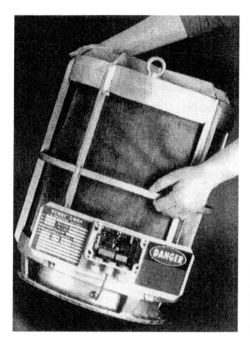

Figure 6-38. Tubular Flame-Quenching Vent System for Flameless Dust Explosion Pressure Relief. (*Courtesy of Rembe GmbH.*)

If vented equipment must be located within buildings, it should be located close to exterior walls so that the vent ducts will be as short as possible, preferably no more than 3 m long.[1]

In the case of dust explosions hazard class KST up to 250 bar x m/s, flame propagation can be prevented altogether by combining the vent panel with a tube-shaped flame trap or flame-quenching vent system as shown in Figure 6-38. By this construction, gases entering the flame trap with a temperature of up to 1500°C (2700°F) leave the flame trap with a temperature of less than 100°C (210°F). This temperature drop reduces the escaping gas volume by a factor of approximately 10, compared with unrestrained venting. All solids are retained by an integrated ceramic filter mat within the tube.

This construction allows explosion venting to be carried out within closed rooms. Passages past the flame trap need only to be shielded against surprise effect.

---

[1] NFPA68 (1988), *Guide for Venting of Deflagrations* (5-4.4, Effects of Vent Ducts).

7

SIZING PRESSURE RELIEF DEVICES

The optimum sizing of pressure relief devices is essential to safely protect a piping system from overpressurization. An undersized system would not be in a position to relieve the system effectively and an oversized system will result in unnecessary material and labor costs.

Flow-through pressure relief devices may be of either one of the following categories. Two of these have subcategories:

- Gas, or vapor, flow
- Subsonic or non-choked flow
- Sonic or choked flow
- Liquid flow
- Mixed-phase flow (vapor plus liquid)
- Gas containing liquid droplets
- Liquid containing gas bubbles
- Liquid flashing to vapor through valve

Each of these has an appropriate formula or procedure for determining the required nozzle size or flow rate. In the case of mixed-phase flow, however, firm recommendations from statutory bodies and standard organizations on appropriate sizing equations are not yet available.[1]

The sizing equations are presented in SI and imperial units. For nomenclature refer to Table 7-1 and Figure 7-1.

---

[1] Compilation of draft document ISO "Design of safety valves and connected vent lines for gas/liquid two-phase flow" in preparation at time of publication of this book.

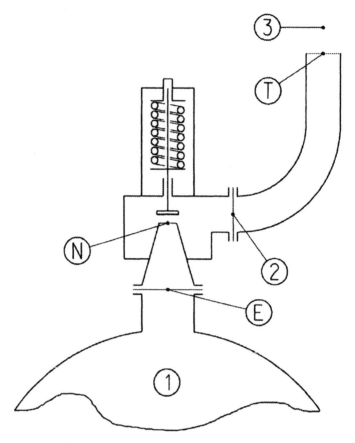

Figure 7-1. Typical Pressure Relief Valve Installation. Subscripts Used in Flow Equation for Conditions at Indicated Points.

Table 7-1
Nomenclature

		Mixed SI Units	Mixed U.S. Units
A	Flow area	mm^2	$in.^2$
C	Gas flow constant for sonic nozzle flow, based on isentropic coefficient k at standard conditions (See Table 7-2. If k is unknown, use $C = 315$ for a conservative result)		
d	Internal pipe diameter	mm	in.

table continued on next page

Table 7-1. (continued)
Nomenclature

		Mixed SI Units	Mixed U.S. Units
f	Pipe friction factor = number of velocity heads lost in length of pipe equal to diameter		
F	Gas flow constant for subsonic nozzle flow, based on isentropic coefficient k at standard conditions and the absolute pressure ratio across the valve		
F_d	Discharge reactive force	N	lb
g	Gravitational acceleration	9.807 m/sec^2	32.174 ft/sec^2
g_c	Proportionality constant	1.0 kgm/Ns^2	$32.174 \text{ lb}_m/\text{ft/lb}_f\text{s}^2$
G	Specific gravity of the liquid referred to water $=1.00$ at $70°F$ $(20°C)$		
k	Isentropic coefficient		
K	Derated or certified coefficient of discharge[1]		
K_b	Capacity correction factor accounting for the effect of built-up and superimposed back pressure on valve lift of bellows balanced pressure relief valves		
K_d	Actual coefficient of discharge[2]		
K_N	Napier capacity correction factor for pressure		
K_{SH}	Napier capacity correction factor for superheat		
K_R	Velocity head loss of rupture devices		
K_v	Viscosity correction factor		
I	Length of pipeline	mm	in.
M	Molecular weight	kg/kmol	lb/lbmol
P	Set pressure (gauge)	barg	$\text{lb/in}^2\text{g}$
P_1	Relieving pressure (gauge) = set pressure plus overpressure plus atmospheric pressure	bara	$1\text{b/in}^2\text{a}$
P_2	Pressure at valve outlet	bara	$\text{lb/in.}^2\text{a}$
P_2'	Pressure in space between valve seat and valve outlet	bara	$\text{lb/in.}^2\text{ a}$
P_3	Atmospheric pressure (abs.)	bara	$\text{lb/in.}^2\text{ a}$

(continued on next page)

[1] ASME (1992) Section VIII Division, UG-131(e)(3).
[2] ASME (1992) Section VIII Division, UG-131(e)(2).

Table 7-1. (continued)
Nomenclature

		Mixed SI Units	Mixed U.S. Units
P_C	Critical vapor pressure, the pressure of saturated vapor at the critical temperature T_C		
P_E	Pressure at valve entry during pressure relief	bara	lb/in.^{2} a
P_N	Pressure at valve nozzle face during pressure relief	bara	lb/in.^{2} a
P_R	Reduced pressure, $P_R = P_1/P_C$		
P_T	Pressure at face of discharge pipe outlet during pressure relief	bara	lb/in.^{2} a
R_0	General gas constant	8314.3 J/kmol°K	1545 ft lb$_f$/lbmol°R
$R_1 = R_0/M$	Individual gas constant	8314.3 J/kg°K	1545 ft lb$_f$/lb°R
Q	Rate of flow of liquid		US gal/min.
T_1	Temperature absolute of gas or vapor at relieving pressure	K = °C + 273	°R = °F + 460
T_C	Critical temperature absolute, the maximum temperature at which the vapor and liquid phases exist in equilibrium		
T_R	Reduced temperature, $T_R = T_1/T_C$		
V	Volumetric flow rate of gas or vapor at standard pressure and temperature	Nm^3/h	scfm
v_1	Specific volume at inlet conditions	m^3/kg	ft^3/lb
W	Mass rate of flow	kg/h	lb/h
Z	Compressibility factor at actual relieving pressure and temperature, correction for deviations from the ideal gas laws		
U	Viscosity Saybolt universal seconds		SU
μ	Dynamic viscosity centipoise	mPa$_1$s	
ρ	Density of liquids	kg/m^3	lb/ft^3

SIZING OF PRESSURE RELIEF VALVES, GAS, VAPOR, STEAM

The equations for the sizing of pressure relief valves for gas or vapor are based on the ideal gas laws in which it is assumed that flow through the

Table 7-2
Values of Critical Pressure Ratios and Coefficients C

k	P_{CF}/P_1	C	k	P_{CF}/P_1	C
1.01	0.60	317	1.45	0.52	360
1.02	0.60	318	1.50	0.51	365
1.04	0.60	320	1.55	0.50	369
1.06	0.59	322	1.60	0.50	373
1.08	0.59	325	1.65	0.49	376
1.10	0.58	327	1.70	0.48	380
1.15	0.57	332	1.75	0.48	384
1.20	0.56	337	1.80	0.47	387
1.25	0.55	342	1.85	0.46	391
1.30	0.55	347	1.90	0.46	394
1.35	0.54	352	1.95	0.45	397
1.40	0.53	356	2.00	0.44	400

valve is isentropic. Deviations from the ideal gas laws are corrected by the compressibility factor Z.

When the ratio of the absolute downstream to the upstream pressure across the nozzle is higher than about 0.5, flow through the valve nozzle is subsonic. At a pressure ratio of about 0.5 and lower, flow through the valve nozzle is sonic. The actual pressure ratio at which flow through the valve nozzle turns sonic is a function of the isentropic coefficient as expressed by the equation

$$\frac{P_{CF}}{P_1} = \left(\frac{2}{K+1}\right)^{k/(k-1)} \tag{7-1}$$

P_{CF} is referred to as the critical nozzle flow pressure and the pressure ratio as the critical pressure ratio. Flow pressure at a pressure ratio below the critical pressure ratio is referred to as subcritical.

Values of the critical pressure ratio for isentropic coefficients 1.01 through 2.00 may be taken from Table 7-2.

Sizing Equations for Gas and Vapor Other Than Steam

Customary Mixed Imperial Units, Sonic Nozzle Flow

Weight flow—lb/h

$$A = \frac{W}{CKP_1 \, K_b} \left(\frac{TZ}{M}\right)^{\frac{1}{2}} \tag{7-2}$$

Volumetric flow—scfm

$$A = \frac{V}{6.32CKP_1\,K_b}(MT_1\,Z)^{\frac{1}{2}} \qquad (7-3)$$

Customary Mixed SI Units, Sonic Nozzle Flow

Weight flow—kg/h

$$A = \frac{131.6W}{CKP_1\,K_b}\left(\frac{TZ}{M}\right)^{\frac{1}{2}} \qquad (7-4)$$

Volumetric flow—Nm³/h

$$A = \frac{5.87V}{CKP_1\,K_b}(MTZ)^{\frac{1}{2}} \qquad (7-5)$$

where

$$C = 520\left[k\left(\frac{2}{k+1}\right)^{\frac{k+1}{k-1}}\right]^{\frac{1}{2}} \qquad (7-6)$$

For values of C, refer to Table 7-2.
For values of K_b, refer to Figure 7-3.

Customary Mixed Imperial Units, Subsonic Nozzle Flow

Weight flow—lb/h

$$A = \frac{W}{735FK_d\,P_1}\left(\frac{TZ}{M}\right)^{\frac{1}{2}} \qquad (7-7)$$

Volumetric flow—scfm

$$A = \frac{V}{4645FK_d P_1}(MTZ)^{\frac{1}{2}} \qquad (7-8)$$

Customary Mixed SI Units, Subsonic Nozzle Flow

Weight flow—kg/h

$$A = \frac{W}{5.58FK_d\,P_1}\left(\frac{TZ}{M}\right)^{\frac{1}{2}} \qquad (7-9)$$

Volumetric flow—scfm

$$A = \frac{V}{125.1 FK_d\, P_1}(MTZ)^{\frac{1}{2}} \qquad (7-10)$$

where

$$F = \left\{\left(\frac{k}{k-1}\right)\left[\left(\frac{P_2'}{P_1}\right)^{\frac{2}{k}} - \left(\frac{P_2'}{P_1}\right)^{\frac{k+1}{k}}\right]\right\}^{\frac{1}{2}} \qquad (7-11)$$

For values of F, refer to Figure 7-2.
Note:

1. The coefficient of discharge as determined at sonic nozzle flow conditions applies also to subsonic nozzle flow, provided the back pressure for determining the value of the factor F is taken as the pressure P_2' in the chamber between nozzle and valve outlet. Valve manufacturers may provide equations that permit the value of P_2' to be calculated from the back pressure P_2 at the outlet flange of the valve.
2. If the back pressure P_2' cannot be determined, a coefficient of discharge should be applied as determined in accordance with API Standard 2000.

Compressibility factor $Z = P_1 V_1/R_1 T_1$. The factor Z compensates for deviations of real gases from the ideal gas laws. Its value is evaluated at inlet conditions. Any correction for Z however is empirical, as the entire derivation of the equations is based on the ideal gas laws.

When critical pressure P_C and critical temperature T_C of the gas are known, the reduced pressure P_1/P_C and the reduced temperature T_1/T_C can be calculated and used to obtain the value of Z from Figures B-1 through Figure B-3 in Appendix B. Values of P_C and T_C for a number of gases may be obtained from Table B-1 in Appendix B. Please note that the critical pressure in Table B-1 is stated in terms of MPa and lb/in^2.

If the compressibility factor cannot be determined, a value of 1.0 is commonly used. This value gives conservative results as long as pressure and temperature are not too high.

Sizing Equations for Dry Saturated Steam

Historically, pressure relief valves for dry saturated steam are sized in the U.S. on Napier's steam flow equation. The equation has been formulated

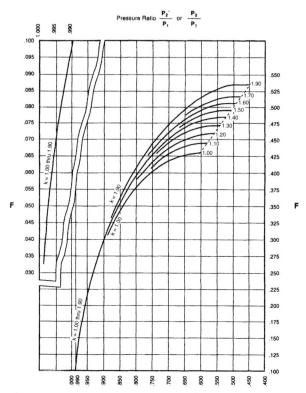

Figure 7-2. Flow Correction Factor F. (*Courtesy of Anderson, Greenwood & Co., Source Catalog 90/9000-US.96.*)

empirically for conditions of choked flow. In the lower pressure range up to 105 bara (1515 psia), Napier's equation gives similar results to sizing equations applicable for any gas, using an isentropic exponent of k = 1.3. At higher pressures, however, Napier's equation can lead to gross oversizing. This led L. Thompson and O. B. Buxton, Jr. to develop a capacity correction factor (K_N) for use in conjunction with Napier's equation.[50]

Customary Mixed Imperial Units

$$A = \frac{W}{51.5 K P_1 \, K_N \, K_{SH}}$$

(7 – 12)

$K_N = 100$ where $P_1 < 1515$ psia

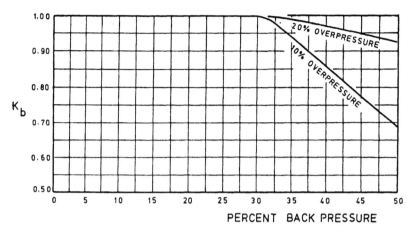

Figure 7-3. Back Pressure Correction Factor K_b, Accounting for the Effect of Built-Up and Superimposed Back Pressure on Valve Lift of Bellows Balanced-Pressure Relief Valves in Gas or Vapor Service. (*Reprinted from API RP 520 Part I, December 1976, Courtesy of American Petroleum Institute.*)

$$K_N = \frac{0.1906P_1 - 1000}{0.2292P_1 - 1061} \quad \text{where } P_1 < 1515 \text{ psia and up to } 3215 \text{ psia}$$

K_{SH} = superheat correction factor Table 7-3.

Values of the correction factor for high-pressure saturated steam, K_N, may be obtained from Figure 7-4.

Customary Mixed SI Units

$$A = \frac{W}{0.525KP_1 \, K_N \, K_{SH}} \tag{7-13}$$

$K_N = 1.00$ where $P_1 < 105$ bara

$$K_N = \frac{2.764P_1 - 1000}{3.323P_1 - 1061} \quad \text{where } P_1 > 105 \text{ bara and up to } 221.7 \text{ bara}$$

K_{SH} = superheat correction factor Table 7-3.

Values of correction factor for high-pressure saturated steam, K_N, may be obtained from Figure 7-4.

Table 7-3
Superheat Correction Factors K_{SH}

Relieving* Pressure			TOTAL TEMPERATURE SUPERHEATED STEAM													
psia	bara	kPaa	400F (204C)	450F (232C)	500F (260C)	550F (288C)	600F (316C)	650F (343C)	700F (371C)	750F (399C)	800F (427C)	850F (454C)	900F (482C)	950F (510C)	1000F (538C)	1050F (566C)
50	3.4	340	.987	.957	.930	.905	.882	.861	.841	.823	.805	.789	.774	.759	.745	.732
100	6.9	690	.998	.963	.935	.909	.885	.864	.843	.825	.807	.790	.775	.760	.746	.733
150	10.3	1030	.984	.970	.940	.913	.888	.866	.846	.826	.808	.792	.776	.761	.747	.733
200	13.8	1380	.979	.977	.945	.917	.892	.869	.848	.828	.810	.793	.777	.762	.748	.734
250	17.2	1720		.972	.951	.921	.895	.871	.850	.830	.812	.794	.778	.763	.749	.735
300	20.7	2070		.968	.957	.926	.898	.874	.852	.832	.813	.796	.780	.764	.750	.736
350	24.1	2410		.968	.963	.930	.902	.877	.854	.834	.815	.797	.781	.765	.750	.736
400	27.6	2760			.963	.935	.906	.880	.857	.836	.816	.798	.782	.766	.751	.737
450	31.0	3100			.961	.940	.909	.883	.859	.838	.818	.800	.783	.767	.752	.738
500	34.5	3450			.961	.946	.914	.886	.862	.840	.820	.801	.784	.768	.753	.739
550	37.9	3790			.962	.952	.918	.889	.864	.842	.822	.803	.785	.769	.754	.740
600	41.4	4140			.964	.958	.922	.892	.867	.844	.823	.804	.787	.770	.755	.740
650	44.8	4480			.968	.958	.927	.896	.869	.846	.825	.806	.788	.771	.756	.741
700	48.3	4830				.958	.931	.899	.872	.848	.827	.807	.789	.772	.757	.742
750	51.7	5170				.958	.936	.903	.875	.850	.828	.809	.790	.774	.758	.743
800	55.2	5520				.960	.942	.906	.878	.852	.830	.810	.792	.774	.759	.744
850	58.6	5860				.962	.947	.910	.880	.855	.832	.812	.793	.776	.760	.744
900	62.1	6210				.965	.953	.914	.883	.857	.834	.813	.794	.777	.760	.745
950	65.5	6550				.969	.958	.918	.886	.860	.836	.815	.796	.778	.761	.746
1000	69.0	6900				.974	.959	.923	.890	.862	.838	.816	.797	.779	.762	.747
1050	72.4	7240					.960	.927	.893	.864	.840	.818	.798	.780	.763	.748
1100	75.6	7580					.962	.931	.896	.867	.842	.820	.800	.781	.764	.749
1150	79.3	7930					.964	.936	.899	.870	.844	.821	.801	.782	.765	.749
1200	82.7	8270					.966	.941	.903	.872	.846	.823	.802	.784	.766	.750
1250	86.2	8620					.969	.946	.906	.875	.848	.825	.804	.785	.767	.751
1300	89.6	8960					.973	.952	.910	.878	.850	.826	.805	.786	.768	.752
1350	93.1	9310					.977	.958	.914	.880	.852	.828	.807	.787	.769	.753
1400	96.5	9650					.982	.963	.918	.883	.854	.830	.808	.788	.770	.754

psia	bara	kPaa										
1500	103.4	10340	.993	.970	.926	.889	.859	.833	.811	.791	.772	.755
1550	106.9	10690		.972	.930	.892	.861	.835	.812	.792	.773	.756
1600	110.3	11030		.973	.934	.894	.863	.836	.813	.792	.774	.756
1650	113.8	11380		.973	.936	.895	.863	.836	.812	.791	.772	.755
1700	117.2	11720		.973	.938	.895	.863	.835	.811	.790	.771	.754
1750	120.7	12070		.974	.940	.896	.862	.835	.810	.789	.770	.752
1800	124.1	12410		.975	.942	.897	.862	.834	.810	.788	.768	.751
1850	127.6	12760		.976	.944	.897	.852	.833	.809	.787	.767	.749
1900	131.0	13100		.977	.946	.898	.862	.832	.807	.785	.766	.748
1950	134.5	13450		.979	.949	.898	.861	.832	.806	.784	.764	.746
2000	137.9	13790		.982	.952	.899	.861	.831	.805	.782	.762	.744
2050	141.3	14130		.985	.954	.899	.860	.830	.804	.781	.761	.742
2100	144.8	14480		.988	.956	.900	.860	.828	.802	.779	.759	.740
2150	148.2	14820			.956	.900	.859	.827	.801	.778	.757	.738
2200	151.7	15170			.955	.901	.859	.826	.799	.776	.755	.736
2250	155.1	15510			.954	.901	.858	.825	.797	.774	.753	.734
2300	158.6	15860			.953	.901	.857	.823	.795	.772	.751	.732
2350	162.0	16200			.952	.902	.856	.822	.794	.769	.746	.729
2400	165.5	16550			.952	.902	.855	.820	.791	.767	.746	.727
2450	168.9	16890			.951	.902	.854	.818	.789	.765	.743	.724
2500	172.4	17240			.951	.902	.852	.816	.787	.762	.740	.721
2550	175.8	17580			.951	.902	.851	.814	.784	.759	.738	.718
2660	179.3	17930			.951	.903	.849	.812	.782	.756	.735	.715
2650	182.7	18270			.952	.903	.848	.809	.779	.754	.731	.712
2700	186.2	18620			.952	.903	.846	.807	.776	.750	.728	.708
2750	189.6	18960			.953	.903	.844	.804	.773	.747	.724	.705
2800	193.1	19310			.956	.903	.842	.801	.769	.743	.721	.701
2850	196.5	19650			.959	.902	.839	.798	.766	.739	.717	.697
2900	200.0	20000			.963	.902	.836	.794	.762	.735	.713	.693
2950	203.4	23040				.902	.834	.790	.758	.731	.708	.688
3000	206.9	20690				.901	.831	.786	.753	.726	.704	.684

*Relieving pressure is the valve set pressure plus the overpressure plus the atmospheric pressure (14.7 psia, 1.0 14 bara or 101.4 kPaa.)

(Table Courtesy of Sempell, Source Engineering Handbook KS 30.593 E)

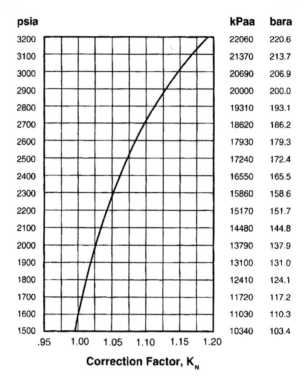

Figure 7-4. Correction Factor for High Pressure Saturated Steam, K_N. (*Courtesy of Crosby Engineering Handbook, Technical Document No. TP-V300, September 1995, Crosby Valve & Gage Company, Wrentham, MA.*)

Sizing Equations for Liquid Flow

Customary Mixed Imperial Units

Volumetric flow—US gpm

$$A = \frac{Q}{38KK_P\,K_V\,K_W} \left(\frac{G}{P_1 - P_2}\right)^{\frac{1}{2}} \qquad (7-14)$$

Customary Mixed SI Units

Weight flow—kg/h

$$A = \frac{0.621W}{KK_P\,K_V\,K_W\,[(P_1 - P_2)\,\rho]^{\frac{1}{2}}} \qquad (7-15)$$

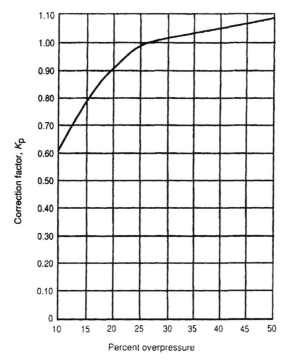

Figure 7-5. Capacity Correction Factor K_p for Pressure Relief Valves in Liquid Service that Requires an Overpressure of 25% to Open Fully. At an Overpressure Lower than 10%, There Is a Heightened Possibility of Valve Chatter. (*Reprinted from API RP 520 Part I, 6th Edition, March 1993, Courtesy of American Petroleum Institute.*)

Note:
K = Coefficient of discharge:

1. For pressure relief valves that open fully on liquid within an overpressure of 10%, obtain the certified value of K from manufacturer.
2. If the pressure relief valve is of the type that requires an overpressure of 25% to open fully, a coefficient of discharge of 0.62 for an overpressure of 25% is commonly assumed. At lower overpressures, a **capacity correction factor KP** to be obtained from Figure 7-5 must be applied. At an overpressure of less than 10%, however, the valve will display a heightened tendency to chatter. For overpressures above 25%, the valve capacity is affected only by the change in overpressure.

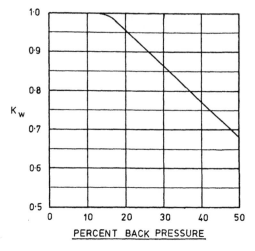

K_w

PERCENT BACK PRESSURE

Figure 7-6. Capacity Correction Factor K_W for Bellows Balanced-Pressure Relief in Liquid Service that Requires an Overpressure of 25% to Open Fully. For Values of K_W for Pressure Relief Valves that in Liquid Service Open Fully within 10% Overpressure, Consult Manufacturer. (Reprinted from API RP 520 Part I, December 1976, Courtesy of American Petroleum Institute.)

K_w = Capacity Correction Factor Due to Back Pressure

The correction factor applies to balanced bellows-type pressure relief valves. Typical values may be obtained from Figure 7-6. For product applicable values, consult valve manufacturer.

Conventional pressure relief valves are affected only by built-up back pressure, though correction for capacity is not commonly allowed for.

K_V = Capacity Correction Factor Due to Viscosity

When a pressure relief valve is to be employed for viscous service, the orifice area has to be determined through an iterative process.

1. Solve initially for non-viscous flow and determine the next available standard orifice size.
2. Determine the Reynolds number, RE, for the initially calculated orifice area from either one of the Equations (7-16) through (7-18).
3. Find value of K_V from Figure 7-7 and apply to the initially calculated discharge area. If the thus corrected initial area exceeds the chosen area, the above calculations should be repeated using the next larger standard orifice size.

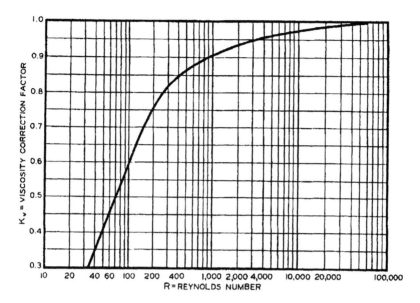

Figure 7-7. Capacity Correction Factor K_V due to Viscosity. (*Reprinted from API RP 520 Part I, December 1976, Courtesy of American Petroleum Institute.*)

Reynolds Number US Units

Volume flow Q = USgpm

$$R_E = \frac{Q2800G}{\mu(A)^{\frac{1}{2}}} \qquad (7-16)$$

$$R_E = \frac{12700Q}{U(A)^{\frac{1}{2}}} \qquad (7-17)$$

Reynolds Number SI Units

Weight flow W = kg/h

$$R_E = \frac{313W}{\mu(A)^{\frac{1}{2}}} \qquad (7-18)$$

where

μ = dynamic viscosity at the flowing temperature, in centipoise
U = viscosity at the flowing temperature, in Saybolt universal seconds

Note: Equation 7-17 is not recommended for viscosities less than 100 Saybolt universal seconds.

Influence of Inlet Pressure Loss on Valve Discharge Capacity

The flow capacity of the discharging valve responds to the pressure at the valve inlet, which is the set pressure minus inlet pressure loss.

In the case of gas flow at sonic velocity, the inlet pressure loss reduces the mass flow rate in direct proportion to the inlet pressure loss. In the case of subsonic gas flow and liquid flow, the change in mass flow depends on the change of pressure difference across the nozzle.

From observation, the industry commonly neglects the capacity loss due to an inlet pressure loss of 3% when sizing direct-acting pressure relief valves, and application codes do not appear to offer guidance at this point. In most practical applications however, the selected nozzle size is bigger than the calculated size by an amount that may compensate partly or fully for the capacity loss due to flow resistance in the inlet pipe.

The same capacity loss due to inlet piping pressure loss is also commonly neglected when sizing pilot-operated pressure relief valves. Any capacity loss due to pressure loss in the inlet piping to the main valve of more than 3% however should be taken into account. In the case of snap-acting flowing-type pilots, the pressure loss in the pipeline from the source of pressure to the pilot connection should in no case exceed 3% of the set pressure to permit stable valve operation at a blowdown of not lower than 5%. Consult manufacturer.

SIZING OF INLET PIPING TO PRESSURE RELIEF VALVES

Codes for the installation of pressure relief valves require that the cumulative total of non-recoverable pressure losses in the inlet piping to the pressure relief valves at rated mass flow, including entry loss at the pressure vessel connection, shall not exceed 3% of the set pressure.

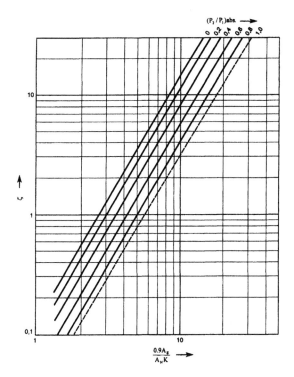

Figure 7-8. Diagram for the Estimation Resistance Coefficient For 3% Pressure Loss in Inlet Line to Pressure Relief Valve, Applicable to Gases and Vapors with an Isentropic Coefficient of k = 1.3. The Dash Line Applies also to Liquids Regardless of Pressure Ratio P_2/P_1. (*Reprinted From AD-Merkblatt A2, Courtesy of Verband der Technischen Uberwachungs-Vereine e.V., Essen.*)

The following procedure permits the estimation of the resistance coefficient of the inlet piping for 3% inlet pressure loss:

1. Initially estimate the resistance coefficient of the proposed inlet piping.
2. Calculate the value of ($0.9A_E/A_N K$). Use this value to obtain from Figure 7-8 the value of the resistance coefficient for an inlet pressure loss of 3%. The isentropic coefficient for gases allowed for in this diagram is 1.3. The dash line applies also to liquids, irrespective of pressure ratio P_2/P_1.
3. If the thus obtained resistance coefficient for 3% inlet pressure loss is equal to or higher than the resistance coefficient of the proposed inlet line, the proposed installation is adequate.

SIZING OF DISCHARGE PIPING OF PRESSURE RELIEF VALVES[1]

Prior to sizing the discharge pipe, obtain from the manufacturer the permissible back pressure P_2 at valve outlet for the type of pressure relief valve under consideration. In the case of bellows balanced pressure relief valves, the permissible back pressure is the sum of built-up and superimposed back pressure.

Initially calculate the resistance coefficient of the proposed discharge pipeline. Then calculate from the equations below the resistance coefficient for the permissible built-up back pressure. If the thus calculated resistance coefficient is equal to or lower than the resistance coefficient of the proposed discharge pipeline, the initially proposed installation is adequate. Otherwise, redesign the discharge pipeline for a lower resistance coefficient.

For liquids:

The built-up back pressure, in terms of the pressure differential across the valve $(P_1 - P_2)$, may be expressed by the Equation 7-19

$$\frac{P_2 - P_3}{P_1 - P_2} = \zeta_2 \left(\frac{KA_N}{0.9A_2} \right)^2 \qquad (7-19)$$

where the pressures P are absolute pressures.

If the permissible back pressure is limited, for example, to 10% of the set pressure, then:

$$\frac{P_2 - P_3}{P_1 - P_2} = 0.10$$

Introducing the above pressure ratio into Equation 7-19 and rearranging the value of the resistance coefficient of the discharge pipeline for 10% built-up back pressure may be obtained from Equation 7-20:

$$\zeta_2 = \frac{0.10}{1 - 0.10} \left(\frac{0.9A_2}{KA_N} \right)^2 \qquad (7-20)$$

---

[1] Sizing procedure taken from Bopp & Reuter. *Safety Valve Handbook far System Planners and Operators.* (*Courtesy of Bopp & Reuther.*)

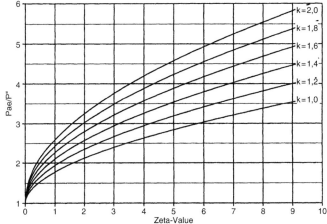

Built-up back pressure Pae in the valve outlet to "critical" pressure P" at the pipe end, dependent on the zeta value of the discharge pipe

Figure 7-9. Diagram for the Estimation of the Resistance Coefficient of the Discharge Pipeline of a Pressure Relief Valve for the Permissible Built-Up Back Pressure in Gas or Vapor Service. (*Courtesy of Bopp & Reuther, "Safety Valves—Handbook for System Planners and Operators."*)

For gas or vapor:

This is the proposed sizing procedure:

1. Calculate the value of the pressure ratio (P_1/P_T), from Equation 7-21

$$\frac{P_T}{P_1} = \left(\frac{2}{k+1}\right)^{\frac{k}{k-1}} \frac{KA_N}{0.9A_T} \qquad (7-21)$$

where the pressures P are absolute pressures.

2. Calculate from the pressure ratio (P_1/P_T) the value of P_T. If the pressure P_T is higher than the atmospheric pressure P_3, the pressure P_T is critical. If the calculated value of P_T is lower than the atmospheric pressure P_3, the pressure at the discharge pipe outlet is the pressure P_3.

3. Calculate the value of P_2/P_T and determine from Figure 7-9 the resistance coefficient of the discharge pipeline for the permissible built-up back pressure in gas or vapor service.

Sizing of Rupture Discs

Rupture discs may be employed in pressure systems subject not only to nonviolent pressure excursions, as in regimes in which pressure relief valves are employed, but also in systems subject to violent pressure excursion resulting from the deflagration of flammable gases or dusts in low-strength containers. Each pressure excursion regime requires a different approach to rupture disc sizing.

Rupture Disc Sizing for Nonviolent Pressure Excursions

The ASME Code (1992)[1] and the API RP 520 Part I (March 1993)[2] present a simplified method of sizing rupture discs by neglecting pressure drops in the inlet piping and in the discharge piping. Based on the equations applicable to pressure relief valves for all fluids, an effective coefficient of discharge of K = 0.62[3] may be used within the following limitations:

Coefficient of discharge method. The coefficient of discharge method, using a coefficient of discharge K equal to 0.62, may be used within the following limitations:

1. The rupture disc discharges directly to the atmosphere, is installed within eight pipe diameters from the vessel nozzle entry, and has a

---

[1] ASME Code (1992), Section VIII, Div. 1, UG-127 (2)(a).

[2] API RP 520, Part 1, 6th edition, 1993, clause 4.8.1.

[3] ISO Standard 6718 and the corresponding European Standard permit the coefficient of discharge of 0.62 for liquid applications only. For compressible fluids, the flow rate is controlled by the nozzle entry configuration of the equipment and the rupture disc device. The standard lists a number of nozzle entry configurations and the applicable coefficient of discharge.

In addition to the coefficient of discharge method, the ISO Standard permits the flow resistance value method, which requires the rupture disc to be sized by analysing the flow resistance of the entire system. This sizing method will be adopted also in the pending revisions of the ASME Code and the API RP 520 Part I, and in the emerging European Standard dealing with the sizing of rupture disc devices. Manufacturers will have to provide the certified flow resistance coefficient of the rupture disc device K_R, which is a dimensionless factor used to calculate the velocity head loss that results from the presence of the rupture disc device in the pressure relief system. This sizing method may require the use of a computer program for speed and accuracy of calculation.

length of discharge pipe not greater than five pipe diameters from the rupture disc device.

2. The nominal diameters of the inlet and discharge piping are equal to or greater than the stamped NPS (or NB) designator of the device.
3. The area, A, to be used in the theoretical flow equation is the minimum net flow area[4] as specified by the rupture disc manufacturer.

Within these limitations, the following sizing equations similar to those for pressure relief valves may be applied:

Sizing Equations for Gas or Vapor Other than Steam

Customary Mixed Imperial Units, Sonic Nozzle Flow

Weight flow—lb/h

$$A = \frac{W}{CK_1P_1}\left(\frac{TZ}{M}\right)^{\frac{1}{2}} \qquad (7-22)$$

Volumetric flow—scfm

$$A = \frac{V}{6.32CKP_1}(MT_1Z)^{\frac{1}{2}} \qquad (7-23)$$

Customary Mixed SI Units, Sonic Nozzle Flow

Weight flow—kg/h

$$A = \frac{131.6W}{CKP_1}\left(\frac{TZ}{M}\right)^{\frac{1}{2}} \qquad (7-24)$$

Volumetric flow—Nm^3/h

$$A = \frac{5.87V}{CKP_1}(MTZ)^{\frac{1}{2}} \qquad (7-25)$$

---

[4] The minimum flow area is the calculated net area after a complete burst of the disc with appropriate allowance for any structural members which may reduce the net flow area through the rupture disc device. The net flow area for sizing purposes shall not exceed the nominal pipe size area of the rupture disc device.

where

$$C = 520 \left[k \left(\frac{2}{k+1} \right)^{\frac{k+1}{k-1}} \right]^{\frac{1}{2}}$$

For values of C, refer to Table 7-2.

Customary Mixed Imperial Units, Subsonic Nozzle Flow

Weight flow—lb/h

$$A = \frac{W}{735 FKP_1} \left(\frac{TZ}{M} \right)^{\frac{1}{2}} \qquad (7-26)$$

Volumetric flow—scfm

$$A = \frac{V}{4645 FKP_1} (MTZ)^{\frac{1}{2}} \qquad (7-27)$$

Customary Mixed SI units—Subsonic Nozzle Flow

Weight flow—kg/h

$$A = \frac{W}{5.58 FKP_1} \left(\frac{TZ}{M} \right)^{\frac{1}{2}} \qquad (7-28)$$

Volumetric flow—Nm³/h

$$A = \frac{V}{125.1 FKP_1} (MTZ)^{\frac{1}{2}} \qquad (7-29)$$

where

$$F = \left\{ \left(\frac{k}{k-1} \right) \left[\left(\frac{P_2}{P_1} \right)^{\frac{2}{k}} - \left(\frac{P_2}{P_1} \right)^{\frac{k+1}{k}} \right] \right\}^{\frac{1}{2}}$$

For Values of F refer to Figure 7-1.

Sizing Equations for Liquid Flow

Customary Mixed Imperial Units

Volumetric flow—US gpm

$$A = \frac{Q}{38KK_V} \left(\frac{G}{P_1 - P_2} \right)^{\frac{1}{2}} \qquad (7-30)$$

Customary Mixed SI Units

Weight flow—kg/h

$$A = \frac{0.621W}{KK_V[(P_1 - P_2)\rho]^{\frac{1}{2}}} \qquad (7-31)$$

Rupture Disc Sizing for Violent Pressure Excursions in Low-Strength Containers

Sizing data on vent panels for explosion venting are based on information from simple models and data obtained from vapor and dust explosions in vessels fitted with explosion relief. The recommended sizing procedures are contained in publications of nationally recognized organizations. The widely recognized publications in the U.S. are the *NFPA Guide for Venting of Deflagrations*, and in Germany *VDI-Richtlinien 3673 (1995)* on Pressure Release of Dust Explosions.[1]

Factors that influence the required size of the explosion vent include:

- Volume and shape of enclosure such as cubic or elongated (silos).
- Homogeneous or non-homogeneous mixture (pneumatic filling).
- Maximum pressure allowed during venting.
- Static bursting pressure of the venting device.
- Explosion severity of the potential hazard, expressed in terms of hazard classes KST for dust or KG for gas.
- Generally, the application of Nomograms.

---

[1] *VDI-Richtlinien 3673 1995* on pressure release of dust explosions are printed in German and English.

8

ACTUATORS

INTRODUCTION

The purpose of this chapter is to introduce the various types of valve actuators currently available to operate the valves previously covered in this book. There are numerous advantages and disadvantages and actuator manufacturer performance data should be referred to in order to assist in the final decision.

A manually operated valve is one that has to be operated by plant personnel, who supply the energy necessary to open, close, or position the closing element of the valve. For a variety of reasons listed below, manual operation may not be an option and an alternative method of valve operation is required. This can be achieved by adding a supplementary piece of equipment called an actuator (Figure 8-1) to the valve. This accessory can, with the energy supplied, operate the valve either by the pressing of a button, or automatically through process systems.

The decision for actuating a valve will be made because of one or more of the following reasons:

- Control of the process system
- Inaccessibility or remote valve location
- Emergency shutdown/fail-safe requirements
- Excessive valve operating torque
- Safety

Figure 8-1. Three-Phase Electric Actuator. (Courtesy of Rotork, UK)

TYPES OF ACTUATORS

The methods of valve actuation fall into one of the following categories:

1. Linear
2. Part-turn
3. Multi-turn

Each of the above can be powered by one of the following power sources:

1. Hydraulic
2. Electric
3. Pneumatic

When choosing the type of actuator the engineer must consider the following factors:

1. The actuator must deliver enough torque to move and seat the valve with the minimum power supply available and also maintain the

required position (open, closed, or intermediate) under the worst flow conditions that the valve might experience.
2. The actuator must be capable of completing the travel, that is, 90°, 180°, and multi-turn at the required speed.
3. The actuator must include a device that prevents excessive forces being applied to the valve under adverse power supply conditions.

VALVE OPERATING FORCES

Linear Valves

These are valves that open and close in a reciprocating manner (i.e., up and down). This can be done with a rotary action on a screwed stem with an electric actuator or directly to a rod, via a cylinder and a pneumatic or hydraulic actuator.

With this type of valve, the actuator has to provide a thrust to overcome surface friction caused between the actuator gland and the valve stem or rod that enters the body.

Gate valves (solid and flexible wedge). These valves shut off the flow by the gate traveling across the bore of the valve. When closing, this will create a throttling effect, causing an increase in the differential pressure across the gate. This effect forces the gate onto the downstream seat and the actuator requires additional thrust in order to overcome this friction.

Parallel slide valves. Similar considerations are required for parallel slide valves, however no additional allowance has to be made for seating as shut-off is achieved by the additional pressure forcing the gate onto the seat.

Globe valves. There are two types of globe valves, the screw down stop and the control valve. The screw down stop type is generally installed so that the process flow is under the disc and therefore resists closing. The globe valve relies on metal-to-metal contact to effect a tight shut-off; it is usual for a high additional thrust allowance to achieve satisfactory sealing between the disc and the seat.

The forces to operate a control valve require more thought, and consideration has to be made for the shape of the closing plug, the flow rate, and the differential pressure.

Diaphragm and pinch valves. The lining material used for this type of valve will vary in stiffness and this has to be considered when calculating the forces required to "pinch" the valve into the closed position. Care must also be taken not to oversize the actuator and therefore deliver too much thrust and rupture the sealing internal membrane.

Part-Turn Valves—90°

Most part-turn valves only have to travel 90° to open or close and have an advantage over multi-turn valves in that because of the smaller distances involved they can be opened or closed using relatively low power consumption.

Ball valves. For ball valves the coefficient of friction between the metal ball and the soft seat or the metal ball and the metal seat has to be considered. Also the effects of the process temperature, pressure, fluid velocity, and viscosity are important factors.

All reputable valve manufacturers supply torque curves, which will illustrate the maximum breakout torque required to initially open the valve from a closed position. This will vary for the numerous types of soft seating materials available, for example PTFE, PEEK, etc. and metal/metal.

Butterfly valves. The velocity of the flow through the valve body has a significant effect on the torque requirements, owing to the hydrodynamic forces generated by the aerofoil type cross section of the valve closure plate. In many cases, actuators for butterfly valves are fitted with self-locking devices, because the closure plate will have a tendency to move because of the reaction forces of the flow through the valve.

Plug valves. Plug valves come in either lubricated or non-lubricated types and the actuator selection will have an effect on the torque requirements to open and close the valve. The design of the plug valve is not pressure-assisted sealing and therefore there is very little difference between the torque requirements between opening and closing the valve. A parallel plug will require less operational torque than a plug with a tapered design.

Conclusion

As mentioned before most valve manufacturers will supply the maximum torque requirements to operate their valves under "normal" service conditions. Additional torque will be required in special cases such as dry gas, slurry, high velocity, high or low temperature, and infrequent use. The manufacturer will also be in a position to specify the media consumption for a full stroke in both directions to open and close.

PNEUMATIC ACTUATORS

Pneumatic actuators are probably the most common type of actuator, because of their cheap and readily available power source, which is compressed air. Not only are they generally less expensive than electric or hydraulic alternatives, but they are also less complex in design and so are easier to maintain.

Pneumatic actuators are particularly suited to part-turn valves with their limited stroke requirements, however they can also be used for linear closing valves. Pneumatic power is relatively easy to store and it is suitable for use when power sources are of a limited capacity or simply not available. Pneumatic actuators can be used on valves of most sizes.

Pneumatic Power Supplies

Pneumatic supply is generally compressed dry air, but designs are also available for natural gas. Compressor and accumulator design is a consideration when designing a network of pneumatically operated actuators to achieve an efficient system. Valve operating times are also a significant factor when determining the pneumatic capacity of the supply and the size of the tubing.

Types of Pneumatic Actuators

There are three basic types of pneumatic actuators, given below.

Diaphragm types. Diaphragm actuators are usually designed for linear motion, although it is possible to have rotary motion designs. Pneumatic actuators are usually attached to the body of the valve by means of a threaded yoke, which can be fitted with a pointer to indicate the position of the valve.

A flexible plate diaphragm is held between two separate casings that are bolted together to form two airtight chambers. There are two independent air supplies to the chambers through which compressed air either can be introduced or can escape. As air is introduced into one chamber, this causes the diaphragm to flex and this motion either extends or retracts the actuator stem, which is attached to the closure element. The more air that is supplied the greater the movement of the stem.

In the single-acting design a spring is introduced to the opposite air chamber to assist the return of the diaphragm when the air pressure is lessened. This spring also acts as a fail-safe mechanism allowing the valve to return to an open or a closed position, when there is a power failure.

Direct acting is where the air pressure introduced forces the diaphragm and the attached valve stem down to the required location. Air failure will cause the stem to be retracted out of the valve body and open the valve, and make sure it remains open.

Reverse acting is where the air pressure introduced forces the diaphragm and attached valve stem up to the required position. Any loss of air pressure will allow the stem to move to the extended position and close the valve. The valve will remain closed until the air supply is reintroduced.

Direct acting for rotary valves is where the air pressure introduced forces the diaphragm and attached valve stem down and, according to its orientation, the valve will be either opened or closed.

Diaphragm valves are relatively cheap, of a basic construction, and easy to maintain. They are very well-suited for low thrust requirements in the lower-pressure piping classes.

Disadvantages of the diaphragm actuators of the linear design are that they can create height problems on smaller valves in restricted locations and on larger sized valves casings can become excessive in weight.

Piston type. Although not initially the first choice of pneumatic design, the piston actuator is becoming more commonly used as process control becomes more sophisticated. This design consists of a cylinder that contains a sliding piston, which is sealed with elastomer O-rings. The actuator cylinder is attached to the valve by a cast yoke that allows for sufficient room on the stem to accommodate positioners and other accessories.

Piston actuators are usually double acting with an air supply to both chambers and a positioner to take the signal and supply or bleed air from the desired air chambers. Pistons generate a linear force that either can be transmitted directly to the actuator stem or has to be converted to a rotary force for use on part-turn valves. As most of the part-turn

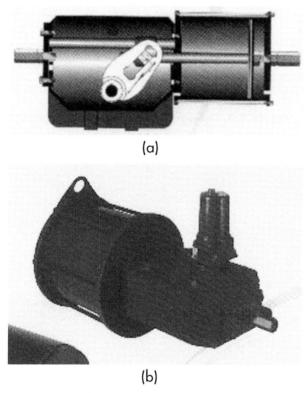

(a)

(b)

Figure 8-2. (a) Scotch Yoke Section; (b) Scotch Yoke. (Courtesy of Rotork, UK)

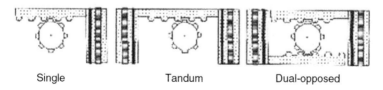

Single   Tandum   Dual-opposed

Figure 8-3. Drawing of Types of Rack and Pinion Actuators.

valves require only 90° movement, this can be achieved by one of the following.

1. A scotch yoke actuator (Figures 8-2a and 8-2b) consists of a piston, a connecting shaft, and a rotary pin.
2. A rack and pinion actuator (Figure 8-3) consists of a single or a double piston that is coupled with an integral rack that drives the pinion.

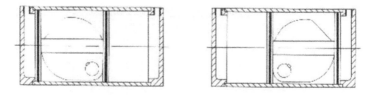

Figure 8-4. Drawing of Double Acting Cam Actuator.

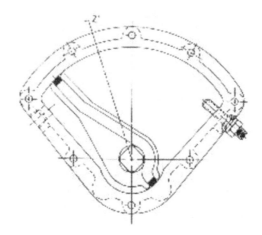

Figure 8-5. Drawing of a Vane Actuator.

3. A trunnion/lever arm actuator normally comprises a simple trunnion mounted cylinder with a piston that acts directly upon a lever that is attached to the valve shaft.
4. A cam actuator (Figure 8-4) consists of double pistons connected by bars with a cam and shaft between them.

Vane type. Vane actuators (Figure 8-5) are commonly used for quarter-turn (90°) applications. They comprise a piston type plate vane, within an airtight chamber, with an integral rotary shaft to produce the necessary torque to rotate the actuator stem. Compressed air is introduced into one of the chambers and air is exhausted from the opposite chamber, causing the vane to move and rotate the shaft.

The advantages of the vane design are low cost, basic design, few moving parts, and low maintenance. Its disadvantage is that, because of its very low torque values, it can only be used for low-pressure applications.

Advantages

To summarize, these are the advantages of pneumatic actuators:

1. Compressed air is convenient and relatively cheap to supply.
2. A piston is not self-locking and it can be used against a spring so it can be used for "single shot" emergency operation.
3. When air is in contact with heat, an increase in pressure caused by thermal expansion will assist the operation of the actuator.
4. Surplus air can be safely vented to the atmosphere.
5. The mechanical design makes it relatively cheaper than alternatives and easier to maintain.

Disadvantages

1. Because there is no kinetic energy it is not possible to deliver an initial "hammerblow" to unseat a wedge valve.
2. Air is a compressible medium and its ability to maintain a fix position is limited. This may result in drifting with both gate and butterfly valves.
3. The design of the pneumatic actuator and the fact that it requires soft sealing rings means that there are temperature limitations and the unit must be equipped with adequate thermal insulation if there is a possibility that the actuator has to function during a fire.
4. Air motors are susceptible to seizure because of ingress of foreign material, or internal corrosion if the compressed air has not been sufficiently dried. This will greatly affect the performance of the actuator.
5. The consumption of air for pneumatic motors will be high and this demand is not always possible for the larger valves requiring higher torques.

Summary

The pneumatic actuator is possibly the most commonly used actuator, because of its advantages and the fact that accessories and spares are readily available. This type of actuator is also available in a variety of metallic materials of construction to suit the application and additional protective coats can be easily applied.

ELECTRIC ACTUATORS

Modern electric motor powered actuators allow local, remote, and hand operation and they are available for a variety of types of valves and sizes. They comprise a revisable electric motor, control box, gearbox, and limit switches. The main advantages of electric actuators are the smooth operation, stability, very high torque values, and maintained thrust, because they are not subject to the problems of a compressible fluid.

Electrical Supply

Motors for multi- and part-turn valves can be either single phase or dc power supplies. However, the three-phase motor is the most commonly used unit, because it is robust and easily controlled.

Environmental Protection

It is imperative that the electric motors used for actuators that will be installed outside and thus exposed to environmental conditions are constructed in such a way that no moisture or condensation can accumulate internally.

Gearing

Electric motors will supply high speeds that will require a large reduction gear ratio to convert to a lower, more realistic speed for valve operation. For part-turn valves, even lower speeds are required, which will require a further reduction.

Manual Operation

In the case of electric motor failure, provision can be made for manual handwheel operation. This handwheel is usually engaged by using a lever clutch that will automatically return to motor driven when the power supply is available.

Advantages

1. Electric motors run at high speeds that can be geared down to suit the required operating torques.

2. Owing to valve gearing, position stability can be maintained.
3. Electric motors can be integrated with process control systems.
4. Electric motors that have been adequately protected from the environment are "clean" and do not generate dirt or moisture.
5. Valves can be easily "hooked up" to the electric power supply.
6. No surplus energy to be dispersed.
7. The electric supply generates a stiff stroke that does not fluctuate.
8. Compact in design.

Disadvantages

1. Expensive.
2. Complex in design and requires higher level of maintenance.
3. Stay put mode when there is a power failure.
4. Electric cables must be protected in a high temperature environment.
5. Electric power is not easily stored.
6. Electrical equipment must be protected from moisture.
7. Electric motors must be intrinsically safe in hazardous areas.

Summary

On the basis of the advantages mentioned above, electric motors are excellent actuators when accurate control of a process system is important. Care must be taken to avoid overestimating the potential power supply and selecting smaller and therefore cheaper units. Electric motors contain fewer moving parts, they require lower maintenance, and generally are energy efficient.

HYDRAULIC ACTUATORS

Hydraulic actuators are capable of delivering very high torques, and with it the fast stroking speed necessary to operate larger valves. They offer stiffness in the stroke, because they are using what is considered to be a non-compressible liquid. Hydraulic actuators are used for these characteristics when, for certain reasons, pneumatic or electric motors are either not suitable or not available.

Basically there are four designs for linear, part-turn, and multi-turn valves:

1. Rack and pinion

2. Hydraulic motor, worm, and gear
3. Rotary piston
4. Piston and cylinder

Hydraulic actuators are available as double acting or spring return, similar to pneumatic actuators and the pressure is supplied by means of a hydraulic pump. This power source can be supplied either locally by a power pack, or remotely with the hydraulic fluid transferred through small bore tubing. The pressure of the hydraulic fluid can be varied to supply different torque outputs. This allows for more flexibility of energy sources and subsequent outputs than is available with both electric and pneumatic actuators.

The energy output of double-acting actuators is constant, however with the spring return type there is a loss of energy during the compression of the spring.

Advantages

1. Higher operating forces can be achieved.
2. Hydraulic pressure can be stored, either locally or remotely.
3. The hydraulic fluid is considered to be incompressible and this results in stability during operation.
4. The hydraulic actuator is compact in size and it can therefore be coated and insulated economically.

Disadvantages

1. No kinetic energy for hammerblow unseating valves.
2. The cost for providing a hydraulic tubing system to supply a network could be economically prohibitive.
3. Remote locations will require long tubing runs and pressure drops that in some cases will prohibit the use of hydraulic powered actuators.
4. Components for hydraulic actuators require close machining tolerances to maintain effective seal characteristics. This means that in the event of fire there is a high probability that these actuators will not operate as required.
5. The hydraulic fluid supply has a relatively high thermal expansion rate and the tubing system will have to be protected from possible fire. Also because of possible thermal expansion, it may be necessary to install a relief device to combat overpressurization.

Summary

Hydraulic and electrohydraulic actuators are robust, very reliable, and they offer a very stiff stroke. The well constructed ones are water-tight and suitable for use in extremely wet environments, such as offshore and sub sea wellheads. However, they are expensive and they can be large and bulky and require special attention during operation.

SIZING ACTUATORS FOR CONTROL VALVES

When an actuator for a control valve is sized it is essential that the worst possible case is considered when the valve is open/closed or in the throttling mode. Control valves that have been incorrectly sized, can result in erratic operation that will lead to vibration and, consequently, wear will be experienced on the closure device and the trim.

When the valve actuator is fully stroked it must overcome any breakout torque and the internal friction experienced by the stem on the various sealing components.

The actuator force available must always be greater than the force available to open/close and throttle the process service efficiently.

ACTUATOR SPECIFICATION SHEET

To ensure that the actuator meets all the process requirements and to avoid any confusion, it is essential that an Engineering Specification Sheet be created for each valve that will be operated by an actuator. This sheet must contain technical information about the actuator and the valve body that it will be attached to. This will avoid any confusion within the design group and it will also allow procurement to confidently place requisitions with potential suppliers.

SPARE PARTS AND MAINTAINING ACTUATED VALVES

When actuators are selected it is usually for the duration of the plant life, which is usually 20 to 25 years. It is not reasonable to expect a complex piece of equipment to survive this long period of time without periodic maintenance and the replacing of certain parts. It is therefore necessary to consider ordering spare parts when the actuator is purchased.

Commissioning Spares

During the commissioning and at the start-up phase of a project, there is a likelihood that the plant will contain dirt and debris in the piping system, even after the lines have been flushed. It is therefore necessary to have a set of soft seals and O-rings in case there is any damage during this pre-production phase.

Two Years' Spares

A budget is created at the beginning of a project for two years' spares, usually while the plant is still under the constructor's warranty. This list of spares is usually more comprehensive than the commissioning spares, because a degree of wear and tear has to be taken into consideration. For actuators this will include soft seals, O-rings, glands, closure discs/plugs, and sets of springs.

Long Term Spares

This list will consider all of the items mentioned for the two years' spares, however serious thought should be given to total replacement of pieces of equipment that are long lead items. Some items are critical for the safe and profitable running of a plant and failure of these parts will prove to be costly.

Maintenance

Part of a manufacturer's responsibility is to deliver a manual that covers the safe operation of the equipment and the maintenance instructions that recommend actions for efficient working. This manual will state the periods between maintenance and what parts require periodic lubrication, greasing, or adjusting.

In the event of major failure during the warranty period, significant work outside of the maintenance regime must be carried out by the manufacturer. Even when a piece of equipment fails outside of its warranty period, the original manufacturer should be the first alternative to be considered.

With major failure it is in many cases cheaper to buy a new piece of equipment, because the cost of the spare parts and the labor often comes very close to the purchase price of a new piece of equipment. Also, with a repair, it will only be a question of time before other parts of the equipment start to fail.

9

DOUBLE BLOCK AND BLEED BALL VALVES

An Introduction to Double Block and Bleed Ball Valves

The increased activity in the offshore sector of the energy industry has led to additional factors that have to be taken into consideration when designing piping systems. Space in these offshore locations is always at a premium and the design of piping systems and their associated components must therefore be more compact. There are structural constraints that are also very important, such as to keep the structure as light as possible, and there are obvious benefits from making components smaller and lighter. Construction labor in offshore locations is also very expensive and any reduction in installation manpower is also beneficial.

The above situation has led to the modification of the patterns of valve components to incorporate savings in space, weight, and labor costs where possible, while still retaining the original function of the valve. Initially these modifications have meant that there is a cost impact, however when installation costs are reduced, the final costs are then comparable with the multi-component option used in the past.

Double Block and Bleed Isolation Philosophy

Process isolation philosophy has become more complex as safety issues have to be addressed and the requirement for double block and bleed isolation has become more commonly used, especially in offshore locations. Double block and bleed isolation requires two in-line isolation valves and a bleed valve, used to drain or vent trapped fluid between the two closure elements. Double block and bleed valves are special components, however there are now several manufacturers of this type of valve. These double block and bleed ball valves have been used very successfully in the offshore sector of the energy industry and their numerous benefits have been realized there not only for new builds, but also for existing plants where they have been retrofitted.

When a process calls for a double block and bleed isolation philosophy, these valves are used for two functions:

1. Instrument double block and bleed (2" and below)—for pressure connections, chemical injection, sampling, vents and drains, etc.
2. In-line double block and bleed (2" and above)—for process isolation.

Instrument Double Block and Bleed Ball Valves

These small bore compact double block and bleed valves (Figures 9-1a and 9-1b) are generally used to isolate instrumentation such as pressure indicators (PIs) and level gauges (LGs) and they come in a variety of combinations. The end connections can be flanged both ends, or flanged one

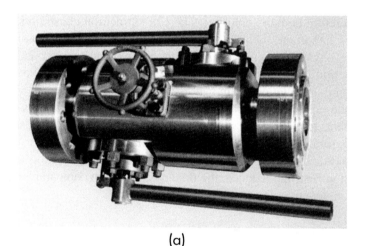

(a)

Figure 9-1. (a) Double Block and Bleed Top Entry. (b) Double Block and Bleed Top Entry General Arrangement. (Courtesy of Orsenigo, Italy)

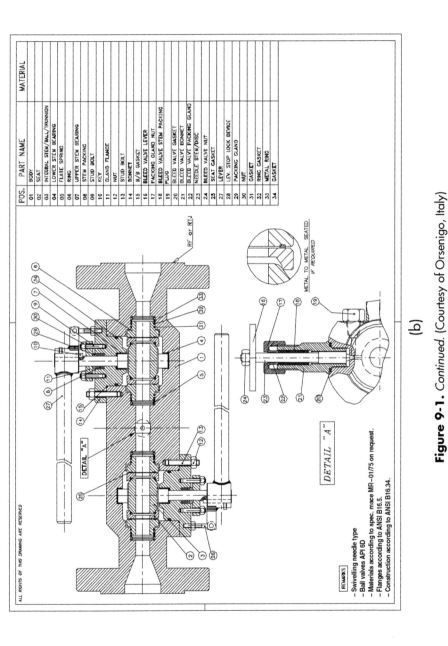

POS.	PART NAME	MATERIAL
01	BODY	
02	SEAT	
03	INTEGRAL STEM/BALL/TRUNNION	
04	LOWER STEM BEARING	
05	PLATE SPRING	
06	RING	
07	UPPER STEM BEARING	
08	STEM PACKING	
09	STUD BOLT	
10	KEY	
11	GLAND FLANGE	
12	NUT	
13	STUD BOLT	
14	BONNET	
15	B/B GASKET	
16	BLEED VALVE LEVER	
17	PACKING GLAND NUT	
18	BLEED VALVE STEM PACKING	
19	PLUG	
20	BLEED VALVE GASKET	
21	BLEED VALVE BONNET	
22	BLEED VALVE PACKING GLAND	
23	NEEDLE STEM/DISC	
24	BLEED VALVE NUT	
25	SEAT GASKET	
27	LEVER	
28	LEV. STOP LOCK DEVICE	
29	PACKING GLAND	
30	NUT	
31	GASKET	
32	RING GASKET	
33	METAL RING	
34	GASKET	

RF or RTJ

METAL TO METAL SEATED IF REQUIRED

DETAIL "A"

REMARKS
- Swivelling needle type
- Ball valves API 6D
- Materials according to spec. mace MR–01/75 on request.
- Flanges according to ANSI B16.5.
- Construction according to ANSI B16.34.

(b)

Figure 9-1. Continued. (Courtesy of Orsenigo, Italy)

end and threaded the other end to receive the instrument gauge. Other end connections such as clamp type are available if specified by the purchaser. The two isolating valves are usually of a quarter-turn ball type, with either soft or metal-metal seats depending on the design conditions of the process fluid. The smaller, usually $\frac{3}{4}"$ or $\frac{1}{2}"$ bleed valve can be a globe, needle, or in some cases a ball type design. The body can be of a one piece or a split type and the ball can be either a floating type or a trunnion type design in the high-pressure classes.

These compact valves can also be used for injecting chemicals or sampling process fluid by adding straight tubing onto the bore of the flange face. When assembled onto a mating flange this quill enters into the process flow and chemicals can be introduced through the valve and then the tube, which is in the process flow. An additional in-line poppet type check valve is incorporated in the compact valve to prevent reverse flow and process fluid contaminating the chemical source. For sampling the reverse is applied and samples can be removed from the process fluid, via the quill, through the open in-line valves and then collected in a sample bottle, which can be taken away for analysis.

(a)

Figure 9-2. (a) In-Line 4" 900 Class Double Block and Bleed Ball Valve Trunnion Mounted Side Entry. (b) General Arrangement Drawing of In-Line Double Block and Bleed. (Courtesy of Orsenigo, Italy)

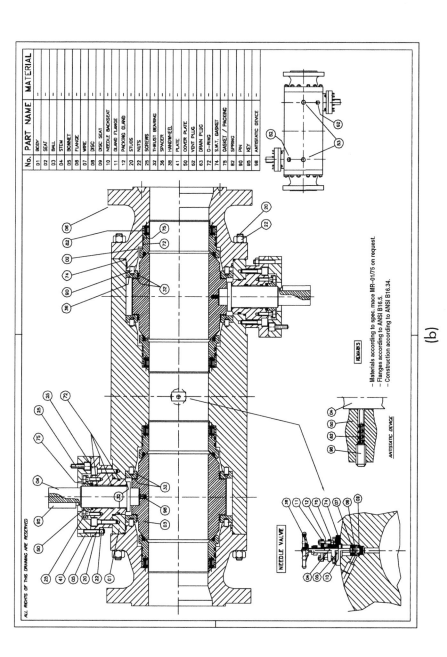

No.	PART NAME	MATERIAL
01	BODY	-
02	SEAT	-
03	BALL	-
04	STEM	-
05	BONNET	-
06	FLANGE	-
07	WIRE	-
08	DISC	-
09	DISC SEAT	-
10	NEEDLE BACKSEAT	-
11	GLAND FLANGE	-
12	PACKING GLAND	-
20	STUDS	-
22	NUTS	-
25	SCREWS	-
32	THRUST BEARING	-
36	SPACER	-
39	HANDWHEEL	-
41	PLATE	-
50	COVER PLATE	-
62	VENT PLUG	-
63	DRAIN PLUG	-
72	O-RING	-
74	S.W.T. GASKET	-
75	GASKET / PACKING	-
82	SPRING	-
90	PIN	-
95	KEY	-
98	ANTISTATIC DEVICE	-

REMARKS

- Materials according to spec. mace MR-01/75 on request.
- Flanges according to ANSI B16.5.
- Construction according to ANSI B16.34.

ANTISTATIC DEVICE

NEEDLE VALVE

(b)

Figure 9-2. *Continued.* (Courtesy of Orsenigo, Italy)

In-Line Double Block and Bleed Ball Valves

These (see Figures 9-2a and 9-2b) are larger in size, usually 2" and above, and they are located in the primary process stream. These compact valves can be supplied in a similar number of end connections as the smaller instrument double block and bleed versions. Because of their size, a threaded connection will rarely if ever be used, however butt-weld end connections are an option that is available, along with clamp type end connections.

These in-line double block and bleed valves are available up to around 8" and pressure ratings up to and including ASME 2500 lb. Availability will depend on the manufacturers and their capabilities. In very special cases some companies are in a position to create special patterns and construct these valves in larger sizes.

In-line double block and bleed ball valves are expensive; however, when you consider the cost of the individual valves, the interconnecting pipework, and the fabrication, the two costs are comparable in sizes up to 8". The double block and bleed valve is a very heavy component and in the larger sizes it will require lifting studies if the valve is to be removed for maintenance.

10

MECHANICAL LOCKING DEVICES FOR VALVES

INTRODUCTION

During recent years more attention has had to be given to the safety of piping systems because of large-scale disasters and the harsher off-shore environments where oil and gas facilities are now constructed. Safer operating environments are required not only to protect profits, but more importantly to protect plant personnel, without whom the former would not be achievable.

Every valve, even the very smallest drain valve, has a role to play in the safe and efficient operation of a process plant. Put in its simplest form a valve can be open, closed, or in a position somewhere in between to control the process flow. All valves are important, however, some valves are naturally more important than others, especially in main process flows and in piping relief systems.

CAR SEALED OPEN AND CAR SEALED CLOSED

During the design stage of a process plant, the Process Engineer will identify some critical valves that must be securely sealed open or sealed closed. These chosen valves are marked on the piping and instrument diagram (P&ID) as CSO or CSC. This stands for Car Sealed Opened and Car Sealed Closed.

This terminology dates back to the very early days of the petroleum industry and it originates from the railroads. Huge quantities of grain and other produce have always been freighted by the railroads all over the United States and there had to be a cheap and efficient method of securing this cargo from theft or illegal passengers boarding the train. The cargo, in most cases, was not particularly valuable and so it was not worth investing vast sums of money to protect it. Also the security system had to be keyless to allow access from various locations along the route of the train and once opened it had to be easily closed again.

The method that was developed was to take one end of a heavy gauge wire, pass it through the car door bolt lock and then through an eyelet on the door frame, and finally twist the two ends together with pliers. The result was cheap and primitive, but it was effective and doors were securely closed and could only be opened by hand with very great difficulty. Railway personnel could open the car doors with the aid of a tool and then close them again quickly. For added security "blobs" of molten lead were added to the ends to guarantee that the sealing system could not be opened by idle hands and that the only method of breaking the seal was with a pair of heavy-duty wire cutters.

Because the railways were used to transport product from the early process plants, this method of temporary securing was adopted by the plant owners. It is still used to this day as the very most basic method for temporarily closing or opening valves.

LOCKED OPEN AND LOCKED CLOSED

As the item that had to be closed is a valve in one location, the reason for having a keyless method of opening was not necessary and the wire was changed to a chain and closed with a padlock. (see Figure 10-1 and Figure 10-2). To change the function of a valve from open to closed or vice versa, can only be done by authorized operators, hence the terminology "key personnel." To this day this is the preferred method of securing valves to be locked open or locked closed. This method prevents the unintentional operation of important valves. Indeed on many of today's projects, valves are often specified with a facility to add a locking device as part of the scope of supply. This could be either a cleat located on the body of the valve, through which a chain or a D-lock can be placed, or a plate with a drilled hole on a ball or plug valve. This method is identified on P&ID drawings and associated plant documentation as LC (locked closed) or LO (locked open).

Figure 10-1. Wire on Gearbox with Independent Padlock Butt Weld. (Courtesy of Netherlocks, Holland)

MECHANICAL INTERLOCKING

As plants grew in size and processes became more complex, operating at higher temperatures and pressures in harsher environments, the need for a more sophisticated method of locking became more important. The possibility of sabotage also had to be considered. These concerns led to the development of the mechanical interlocking system, an engineered piece of equipment that can be fitted to all manual and actuated valves and only allows operation to be carried out in a sequence with other valves.

The mechanical locking device is essentially two components: first a metal housing with integral locks that is attached by bolts to the top works of the valve with the operator, either a handwheel or a wrench. Secondly there are two keys dedicated to these specific locks for opening or closure. One key is used to lock open the valve and the other key is used to lock closed the valve. The loose key, the one not in the valve, is always retained in a key cabinet (see Figures 10-3a and 10-3b) that is located in the control

Figure 10-2. Wire on Globe Valve Handwheel with Independent Padlock. (Courtesy of Netherlocks, Holland)

room, or used to open/close another valve. Only authorized "key" personnel will have access to this cabinet.

Mechanical locking is a relatively inexpensive method of aiding process control and adding to the safe operation of a plant for the benefit of the plant personnel and the owner. These mechanical devices can be either installed during the construction phase of the plant or retrofitted while the plant is operating. At the moment mechanical locking devices are the best solution to creating a safe operating environment.

These locking devices are of a robust design and constructed of a combination of materials that both have mechanical strength and are also suitable for the environmental conditions. Typically the housing will be constructed of a combination of carbon steel and stainless steel, however the key is most likely to be made of stainless-steel plate (see Figure 10-4, Figure 10-5, and Figure 10-6). Valves that are locked open or locked closed are usually identified as LO (locked open) or LC (locked closed); in addition to this an SP (Special Piping) is called up indicating that a special piece of equipment, that is mechanical interlocking, is required at a specific location.

(a)

(b)

Figure 10-3. (a) and (b) Key Cabinets. (Courtesy of Netherlocks, Holland)

Figure 10-4. Gate Valve with Mechanical Locking Device. (Courtesy of Netherlocks, Holland)

Figure 10-5. Ball Valve with Mechanical Locking Device. (Courtesy of Netherlocks, Holland)

Figure 10-6. Butterfly Valve with Mechanical Locking Device. (Courtesy of Netherlocks, Holland)

THE MECHANICAL INTERLOCKING OF PRESSURE SAFETY VALVES

Process plants that have pressurized piping systems must also have an integrated pressure relief system to avoid explosions caused by over-pressurization and possible in-service pipe failure. Pressure safety valves (PSVs) are the barrier between on the upstream the pressurized system and on the downstream relief to the atmosphere via an unvalved header.

A PSV contains a spring that is set to "pop" when overpressurization of the process piping system occurs and fluid must be released to the atmosphere to avoid component failure. In order to execute maintenance to the in-service live PSV, a spare PSV of the same specification must be available for switchover. This change-over can take place while the plant is live, if the correct piping configuration is laid out. When this maintenance activity takes place, there is always a risk that the piping system or the vessel is left unprotected, which is a safety threat in the event of overpressurization. To avoid this risk, mechanical interlocks can be installed on the upstream

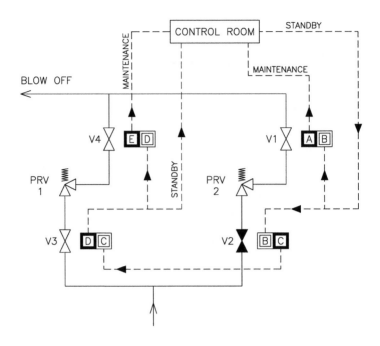

Figure 10-7. Double PRV Schematic (Courtesy of Netherlocks, Holland)

and downstream isolating valves associated with the PRV. With these inter-locks installed the operators will only be able to open or close valves in a specific sequence, which means that the piping system or the vessel is always protected by a minimum of one PSV.

Procedure to Change Out PRV 1

See Figure 10-7.

Step 1—Starting scenario

PRV 1 in service
PRV 2 spare
V1, V3, V4—locked open
V2—locked closed
Key B in the control room

Step 1. Key B is collected from the control room and inserted into the empty slot on V2. This valve can now be opened and locked and release Key C.

Step 2. Key C can be inserted into the slot of V3, releasing Key D and allowing the valve to be locked closed.

Step 3. Key D is used to lock closed V4 and release Key E.

Step 4. Return Key E to the control room.

Intermediate scenario

PRV 1 out of service to be removed for repair
PRV 2 in service
V1, V2—locked open
V3, V4—locked closed
Key E in the control room

PRV 1 can be repaired and re-installed and Key E can be taken from the control room and used to lock open V4. This releases Key D which is returned to the control room.

Final scenario

PRV 1 spare
PRV 2 in service
V1, V2, V4—locked open
V3—locked closed
Key D in the Control Room

THE MECHANICAL INTERLOCKING OF PIPELINE LAUNCHERS AND RECEIVERS

Pipelines carrying product require periodic cleaning and this is achieved by cleaning elements called "pigs," which are propelled through the pipeline by water or the service fluid. These pigs are introduced into the pipeline via a launcher, which is a piece of equipment comprising a pipe barrel, a closure door, and various nozzles.

When the pig is introduced into the pipeline it is essential that full depressurization of the piping system has taken place, thus avoiding any accidents when the closure door has been opened while the system has been live. There have been numerous fatalities and injuries when this has happened and by introducing a mechanical interlocking procedure this problem can be solved.

Normal Operation Conditions

Under normal operating conditions the pipeline pig launcher is isolated and all valves are locked closed.

Starting scenario. See Figure 10-8.

>Vent valve V4—locked closed
>Isolating valve V3—locked closed
>Pipeline valve V2—locked open
>Isolating valve V1—locked closed
>Launcher closure door—locked closed
>Key A in the control room

>**Step 1.** Key A is removed from the control room and used to open the vent valve V4. This releases Key B.
>**Step 2.** Key B is now used to open the closure door at the end of the launcher barrel.

Loading the pig

>Vent valve V4—locked open
>Isolating valve V3—locked closed
>Pipeline valve V2—locked open
>Throttling valve V1—locked open
>Launcher closure door—locked open
>No key in the control room

>The pig can be placed in the barrel and prepared for launching.

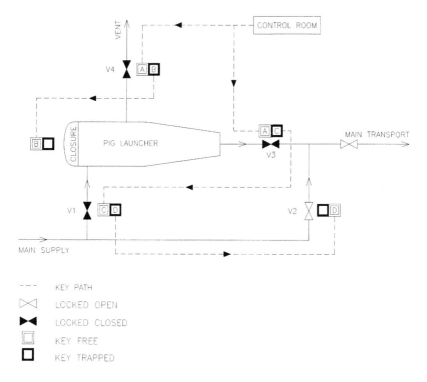

Figure 10-8. Pig Launcher Schematic. (Courtesy of Netherlocks, Holland)

Launching the pig

To re-instate the launcher, first a reverse key sequence has to take place.

Step 3. The launcher door is locked closed which releases Key B, which is used to lock closed vent V4. This releases Key A.

Step 4. Key A is used to lock open the isolating valve V3, which releases Key C.

Step 5. Key C is used to lock open valve V1, which releases Key D.

At this point the pig launcher is under full line pressure.

Step 8. Key D is used to lock closed the main pipeline valve V2.

The throttle valve, V2, is then used to restrict the flow through the launcher, which causes a pressure build-up behind the pig, until friction is overcome causing the pig to start its travel along the pipeline. Pressure to drive the pig will be sustained until the pig has completed its journey and arrived at the receiver vessel. The receiver will also be protected by a similar mechanically interlocked set of valves and a sequence of opening and closing valves will have to be carried out before the receiver is depressurized, the door can be opened, and the pig retrieved from the barrel.

What has been described above is the basic procedure to launch and receive a pig, and this activity can be more complex with higher pressures, different piping configurations, and the addition of PSVs. But the objective is the same—that the launcher or a receiver can only be opened when they have been safely depressurized and drained of excess fluid.

CONCLUSION

The use of mechanically interlocked valves has worked very successfully in recent years and because of the increase in safety, they have become a valuable addition to petrochemical plant equipment. There have been numerous plant accidents in the past that could have been avoided if an interlock system had been in place. Mechanical interlocks should not be looked upon as an accessory, but as an integral part of a piping safety relief system, which is essential to guarantee safe plant operation.

Appendix A

ABBREVIATIONS OF ORGANIZATIONS AND STANDARDS RELATED TO THE INTERNATIONAL VALVE INDUSTRY

Abbreviation	Name of Organization
ACI	Alloy Casting Institute
AFNOR	Association Francaise de Normalisation (France)
AIChE	American Institute of Chemical Engineers
AISI	American Iron and Steel Institute
ANSI	American National Standards Institute
API	American Petroleum Institute
ASM	American Society of Metals
ASME	American Society of Mechanical Engineers
ASTM	American Society for Testing and Materials
AWS	American Welding Society
BSI	British Standards Institute
CGA	Compressed Gas Association
CSA	Canadian Standards Association
DIN	Deutsches Institut fur Normung (Germany)
EPA	Environmental Protection Agency
FDA	Food and Drug Administration
HSE	Health and Safety Executive
ISA	Instrument Association of America

ISO	International Standards Organization
JIS	Japan Industrial Standard
MSS	Manufacturers Standardization Society
NACE	National Association of Corrosion Engineers
NFPA	National Fire Prevention Association
OCMA	Oil Companies Materials Association
PFI	Pipe Fabrication Institute
PIMA	Paper Industry Management Association
PLCA	Pipe Line Contractors Association
SAE	Society of Automotive Engineers
SPE	Society of Petroleum Engineers
TEMA	Tubular Exchanger Manufacturers Association
VMA	Valve Manufacturers Association

Appendix B

PROPERTIES OF FLUIDS

This section contains the following:

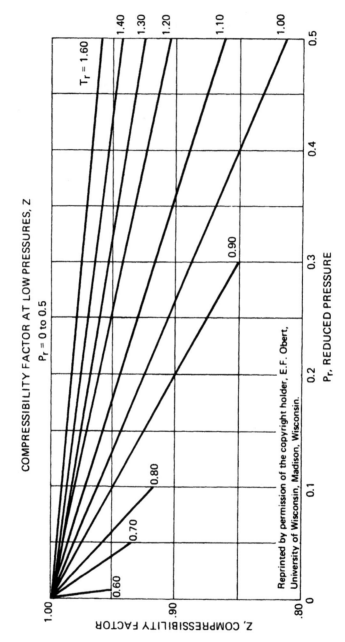

Figure B-1. Compressibility factor, Z, at low pressures for P_r = 0 to 0.5.

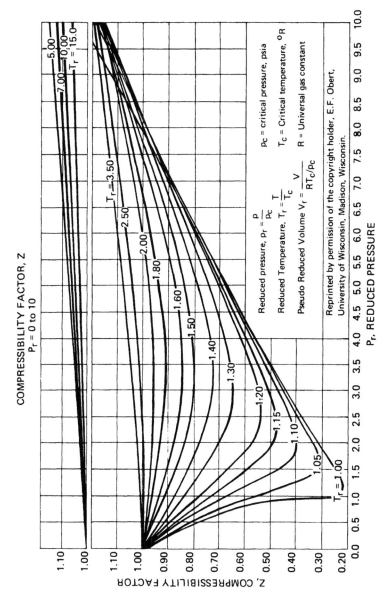

Figure B-2. Compressibility factor, Z, for $P_r = 0$ to 10.

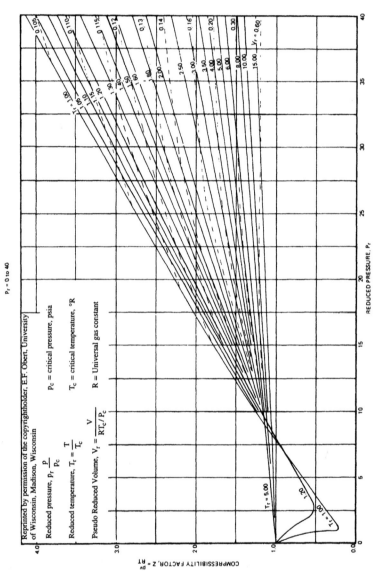

Figure B-3. Compressibility Factor, Z_1 for $P_r = 0$ to 40.

Physical Constants of Gases

Gas		M Molecular Mass		R Individual Gas constant		Isentropic Coefficient k for P → atm T = 273K = 492R	Tc Critical Temperature		Pc Critical Pressure (absolute)	
		kg/kmol	lb/lbmol	J/kg.K	ft.lbf/lb.°R		°K	°R	MPa	lb/in²
Acetylene	C_2H_2	26.078		318.82	59.24	1.23	309.09	556.4	6.237	904.4
Air	—	28.96		287.09	53.35	1.40	132.4	238.3	3.776	547.5
Ammonia	NH_3	17.032		488.15	90.71	1.31	405.6	730.1	11.298	1638.2
Argon	Ar	39.994		208.15	38.68	1.65	150.8	271.4	4.864	705.3
Benzol	C_6H_6	78.108		106.44	19.78	—	561.8	1011.2	4.854	703.9
Butane-n	C_4H_{10}	58.124		143.04	26.58	—	425.2	765.4	3.506	508.4
Butan-i	C_4H_{10}	58.124		143.04	26.58	—	408.13	734.6	3.648	529.0
Butylene	C_4H_8	56.108		148.18	27.54	—	419.55	755.2	3.926	569.2
Carbon dioxide	CO_2	44.011		188.91	35.10	1.30	304.2	547.6	7.385	1070.8
Carbon disulfide	CS_2	76.142		109.19	20.29	—	546.3	983.3	7.375	1069.4
Carbon monoxide	CO	28.011		296.82	55.16	1.40	133.0	239.4	3.491	506.2
Carbon oxysulfide	COS	60.077		138.39	25.72	—	375.35	675.6	6.178	895.9
Chlorine	Cl_2	70.914		117.24	21.79	1.34	417.2	751.0	7.698	1116.3
Cyanogen	C_2N_2	52.038		159.77	29.69	—	399.7	719.5	5.894	854.6
Ethane	C_2H_6	30.070		276.49	51.38	1.20	305.42	549.8	4.884	708.2
Ethylene	C_2H_4	28.054		296.36	55.07	1.25	282.4	508.3	5.070	735.2
Helium	He	4.003		2076.96	385.95	1.63	5.2	9.4	0.229	33.22
Hydrogen	H_2	2.016		4124.11	766.36	1.41	33.3	59.9	1.295	187.7
Hydrogen chloride	HCl	36.465		228.01	42.37	1.39	324.7	584.5	8.307	1204.4
Hydrogen cyonide	HCN	27.027		307.63	57.16	—	456.7	822.1	5.394	782.1
Hydrogen sulfide	H_2S	34.082		243.94	45.33	1.33	373.53	672.4	9.013	1306.8

(table continued)

Table B-1. *(continued)*
Physical constants of Gases

Gas		M Molecular Mass (kg/kmol)	M Molecular Mass (lb/lbmol)	R Individual Gas constant (J/kg.K)	R Individual Gas constant (ft.lbf/lb.°R)	Isentropic Coefficient k for P → atm T = 273K = 492R	T_c Critical Temperature (°K)	T_c Critical Temperature (°R)	P_c Critical Pressure (absolute) (MPa)	P_c Critical Pressure (absolute) (lb/in²)
Methane	CH_4	16.034		518.24	96.30	1.31	190.7	343.3	4.629	671.2
Methyl chloride	CH_3Cl	50.491		164.66	30.60	—	416.2	749.2	6.669	667.0
Neon	Ne	20.183		411.94	76.55	1.64	44.4	79.9	2.654	384.8
Nitric oxide	NO	30.008		277.06	51.48	1.39	180.2	324.4	6.541	948.5
Nitrogen	N_2	28.016		296.76	55.15	1.40	126.3	227.3	3.383	490.6
Nitrous oxide	N_2O	44.016		188.89	35.10	1.28	309.7	557.5	7.267	1053.7
Oxygen	O_2	32.000		259.82	48.28	1.40	154.77	278.6	5.080	736.6
Sulfur dioxide	SO_2	64.066		125.77	24.11	1.28	430.7	775.3	7.885	1143.3
Propane	C_3H_8	44.097		188.54	35.04	—	370.0	666.0	4.256	617.15
Propylene	C_3H_6	42.081		197.56	36.71	—	364.91	656.8	4.621	670.1
Toluene	C_7H_8	92.134		90.24	16.77	—	593.8	1068.8	4.207	610.0
Water vapor	H_2O	18.016		461.48	85.75	1.33*	647.3	1165.1	22.129	3208.8
Xylene	C_8H_{10}	106.16		78.32	14.55	—	—	—	—	—

*at 100°C.

Source: VD Blatt 4, Entwurf, Jan. 1970, Berechnungsgrundlagen für die Durchflussmessung mit Drosselgeräten, Stoffwerte.
By Courtesy of VDI/VDE—Gesellschaft für Mess-und Regelungstechnik.

Appendix C

STANDARDS
PERTAINING TO VALVES

This chapter lists common USA and British standards pertaining to valves, as published in the standard indexes of the various standard organizations for 2003. Because new standards are continually issued and old standards revised or withdrawn, the validity of these standards should be verified prior to application.

STANDARD ORGANIZATIONS

ANSI   American National Standards Institute
         1819 L Street, N.W.
         New York, New York 10018
         Telephone: +1-202-293-8020, Fax: +1-202-293-9287
         email: info@asme.org, website: www.asme.org
API   American Petroleum Institute
         1220 L Street, N.W.
         Washington, D.C. 20005-4070
         Telephone: +1-202-682-8000, Fax: +1-202-682-8408
         email: info@api.org, website: www.api.org
ASME   The American Society of Mechanical Engineers
         Three Park Avenue
         New York, New York 10016-5990
         Telephone: +1-973-882-1167
         email: infocentral@asme.org, website: www.asme.org

ASTM   ASTM International
       100 Barr Harbour Drive
       West Conhohcken, PA 19428-2959
       Telephone: +1-610-832-9585, Fax: +1-610-832-9555
       email: info@astm.org, website: www.astm.org
AWWA   American Water Works Association
       6666 West Quincy Avenue
       Denver, CO 80235
       Telephone: +1-303-794-7711, Fax: +1-303-347-0804
       email: info@awwa.org, website: www.awwa.org
MSS   Manufacturers Standardization Society
       127 Park Street N.E.
       Vienna, VA 22180-4602
       Telephone: +1-703-281-6613, Fax: +1+703-281-6671
       email: info@mss-hq.com, website: www.mss.hq.com
AFNOR Association Francaise de Normalisation
       11, avenue Francis de Pressense
       FR-93571 Saint Denis la Plaine Cedex
       France
       Telephone: +33-1-41-62-80-00, Fax: +33-1-49-17-90-00
       email: uari@afnor.fr, website: www.afnor.fr
BSI   British Standards Institute
       389 Chiswick High Road
       London W4 4AL
       United Kingdom
       Telephone: +44-(0)208-996-9000, Fax: +44-(0)208-996-7001
       email: info@bsi-global.com, website: www.bsi-co.uk
DIN   Deutsches Institut fur Normung eV
       Burggrafenstrasse 6
       10787 Berlin
       Germany
       Telephone: +49-30-2601-0, Fax: +49-30-2601-1260
       email: postmaster@din.de, website: www.din.de

STANDARDS PERTAINING TO VALVE ENDS AND GENERAL VALVE STANDARDS

MSS SP-6   Standard finishes for contact faces of pipe flanges and connecting-end flanges of valves and fittings.

MSS SP-9	Spot facing for bronze, iron, and steel flanges.
MSS SP-44	Steel pipeline flanges.
MSS SP-51	Class 150 LW corrosion-resistant cast flanges and flanged fittings.
MSS SP-65	High-pressure chemical industry flanges and threaded stubs for use with lens gaskets.
MSS SP-91	Guidelines for manual operation of valves.
MSS SP-92	MSS valve user guide.
MSS SP-96	Guidelines on terminology for valves and fittings.
MSS SP-98	Protective coatings for the interior of valves, hydrants, and fittings.
MSS SP-99	Instrument valves.
MSS SP 120	Flexible graphite packing systems for rising stem steel valves.
API Spec 6A	Wellhead and Christmas tree equipment.
API Std 605	Large diameter carbon steel flanges.
ASME B1.20.1	Pipe threads, general purpose (inch).
ASME B1.20.3	Dryseal pipe threads (inch).
ASME B16.1	Cast iron pipe flanges and flanged fittings.
ASME B16.5	Pipe flanges and flanged fittings.
ASME B16.20	Metallic gaskets for pipe flanges.
ASME B16.21	Non-metallic flat gaskets for pipe flanges.
ASME B16.24	Cast copper pipe flanges and flanged fittings.
ASME B16.25	Butt-welding ends.
ASME/AWWA C207-78	Flanges for water-works service, 4 in. through 144 in. steel.
ASME/AWWA C606-78	Joints, grooved and shouldered type.
BS 21	Pipe threads for tubes and fittings where pressure-tight joints are made on the threads (metric dimensions).
BS 1560	Steel pipe flanges and flanged fittings (nominal sizes $\frac{1}{2}$ in. to 24 in.) for the petroleum industry. Part 2 (1970), metric dimensions.
BS 3293	Carbon steel flanges (over 24 in. nominal size) for the petroleum industry.
BS 4504	Flanges and bolting for pipes, valves, and fittings, metric series. Part 1 (1969), ferrous. Part 2 (1974), copper alloy and composite flanges.

STANDARDS PERTAINING TO GLOBE VALVES

MSS SP-42	Class 150 corrosion-resistant gate, globe, angle, and check valves with flanged and butt-weld ends.
MSS SP-61	Pressure testing of steel valves.
MSS SP-80	Bronze gate, globe, angle, and check valves.
MSS SP-85	Gray iron globe and angle valves, flanged and threaded ends.
MSS SP-117	Bellows seals for globe and gate valves.
MSS SP-118	Compact steel globe and check valves—flanged, flangeless, threaded and welding ends.
API RP 6FA	Fire test for valves.
ASME B16.10	Face-to face and end-to-end dimension of ferrous valves.
ASME B16.34	Steel valves, flanged and butt-welding end.
BS 1873	Steel globe valves and stop and check valves (flanged and butt-welding ends), for the petroleum, petrochemical, and allied industries.
BS 5352	Cast and forged steel wedge gate, globe, check, and plug valves, screwed and socket welding, sizes 50 mm and smaller, for the petroleum, petrochemical, and allied industries.
BS 5152	Cast iron globe and globe stop and check valves, for general purposes.
BS 5154	Copper alloy globe, globe stop and check, check, and gate valves (including parallel slide type), for general purposes.
BS 5160	Specification for flanged steel globe valves, globe stop and check valves, and lift-type check valves for general purposes.

STANDARDS PERTAINING TO PARALLEL AND WEDGE GATE VALVES

MSS SP-42	Class 150 corrosion-resistant gate, globe, angle, and check valves with flanged and butt-weld ends.
MSS SP-45	Bypass and drain connection standard.
MSS SP-61	Pressure testing of valves.
MSS SP-70	Cast iron gate valves, flanged and threaded ends.

MSS SP-80	Bronze gate, globe, angle, and check valves.
MSS SP-81	Stainless steel, bonnetless, flanged, wafer, knife gate valves.
MSS SP-117	Bellows seals for globe and gate valves.
API Spec 6D	Specification for pipeline valves, end closures, connectors and swivels.
API RP 6FA	Fire test for valves.
API Std 595	Cast iron gate valves, flanged ends.
API Std 597	Steel venturi gate valves, flanged or butt-welding ends.
API Std 598	Valve inspection and test.
API Std 600	Bolted bonnet steel gate valves for petroleum and natural gas industries.
API Std 602	Compact carbon steel gate valves.
API Std 603	Corrosion-resistant, bolted bonnet gate valves— flanged and butt-welding ends.
ASME B16.10	Face-to-face and end-to-end dimensions of ferrous valves.
ASME B16.34	Steel valves, flanged and butt-welding end.
BS 1414	Steel wedge gate valves (flanged and butt-welding ends) for the petroleum, petrochemical and allied industries.
BS 5150	Cast iron wedge and double-disc gate valves, for general purposes.
BS 5154	Copper alloy globe, globe stop and check, check, and gate valves (including parallel slide type), for general purposes.
BS 5163	Double-flanged cast iron wedge gate valves for water works purposes.
BS 5352	Cast and forged steel wedge gate, globe, check, and plug valves, screwed and socket welding, sizes 50 mm and smaller, for the petroleum, petrochemical, and allied industries.

STANDARDS PERTAINING TO PLUG VALVES

MSS SP-61	Pressure testing of steel valves.
MSS SP-78	Cast iron plug valves, flanged and threaded ends.

MSS SP-108	Resilient-seated cast iron-eccentric plug valves.
API Spec 6A	Wellhead and Christmas tree equipment.
API Spec 6D	Pipeline valves (gate, plug, ball, and check valves).
API RP 6FA	Fire test for valves.
API Std 599	Metal plug valve—flanged and welding ends.
ASME B16.10	Face-to-face and end-to-end dimensions of ferrous valves.
ASME B16.34	Steel valves, flanged and butt-welding end.
BS 5158	Cast iron and cast steel plug valves for general purposes.
BS 5353	Specification for plug valves.

STANDARDS PERTAINING TO BALL VALVES

MSS SP-61	Pressure testing of steel valves.
MSS SP-68	High pressure butterfly valves with offset design.
MSS SP-72	Ball valves with flanged or butt-welding ends for general service.
MSS SP-110	Ball valves threaded, socket-welding, solder joint, grooved and flared ends.
MSS SP-122	Plastic industrial ball valves.
API Spec 6D	Pipeline valves (gate, plug, ball, and check valves).
API Std 598	Valve inspection and test.
API Std 607	Fire test for soft-seated quarter-turn valves.
ASME B16.10	Face-to-face and end-to-end dimensions of ferrous valves.
ASME B16.34	Steel valves, flanged and butt-welding end.
BS 5159	Cast iron carbon steel ball valves for general purposes.
BS 5351	Steel ball valves for the petroleum, petrochemical, and allied industries.

STANDARDS PERTAINING TO BUTTERFLY VALVES

MSS SP-67	Butterfly valves.
API Std 598	Valve inspection and test.
API Std 609	Butterfly valves, "double flanged" lug-type and wafer-type.
ANSI/AWWA C504-80	Rubber-seated butterfly valves.

STANDARDS PERTAINING TO DIAPHRAGM VALVES

MSS SP-88	Diaphragm-type valves.
BS 5156	Screwdown diaphragm valves for general purposes.
BS 5418	Marking of general purpose industrial valves.
ISO 5209	Marking of general purpose industrial valves.
DIN 3359	*Membran-Absperrarmaturen aus metallischen Werkstoffen.*

STANDARDS PERTAINING TO STAINLESS STEEL VALVES

MSS SP-42	Class 150 corrosion-resistant gate, globe, angle, and check valves with flanged and butt-weld ends.
API Std 603	Corrosion-resistant, bolted bonnet gate valves.

STANDARDS PERTAINING TO CHECK VALVES

MSS SP-42	Class 150 corrosion-resistant gate, globe, angle, and check valves with flanged and butt-weld ends.
MSS SP-61	Pressure testing of steel valves.
MSS SP-71	Gray iron swing check valves, flanged and threaded ends.
MSS SP-80	Bronze gate, globe, angle, and check valves.
MSS SP-118	Compact steel globe and check valves—flanged, flangless, threaded and welding ends.
MSS SP-126	Steel in-line spring assisted center guided check valves.
API Spec 6D	Pipeline valves (gate, plug, ball and check valves).
API RP 6FA	Fire test for valves.
API Std 594	Check valves: wafer, wafer-lug and double flanged type.
ASME B16.10	Face-to-face and end-to-end dimensions of ferrous valves.
ASME B16.34	Steel valves, flanged and butt-welding end.
BS 1868	Steel check valves (flanged and butt-welding ends) for the petroleum, petrochemical, and allied industries.

BS 1873	Steel globe and glove stop and check valves (flanged and butt-welding ends) for the petroleum, petrochemical, and allied industries.
BS 5152	Cast iron globe and globe stop and check valves for general purposes.
BS 5154	Copper alloy globe, globe stop and check, check, and gate valves.
BS 5160	Specification for flanged steel globe valves, globe stop and check valves, and lift-type check valves for general purposes.
BS 5352	Cast and forged steel wedge gate, globe, check, and plug valves, screwed and socket-welding, sizes 50 mm and smaller, for the petroleum, petrochemical, and allied industries.

STANDARDS PERTAINING TO PRESSURE VALVES

API RP 520	Recommended practice for the design and installation of pressure relieving systems in rcfineries.
	Part I (1976)—Design
	Part-II (1973)—Installation.
API RP 521	Guide for pressure relief and depressurizing systems.
ASME/API 526	Flanged-steel safety relief valves.
ASME/API 527	Commercial seat tightness of safety relief valves with metal-to-metal seats.
BS 6759	Safety valves.
Part 1	Safety valves for steam and hot water.
Part 2	Safety valves for compressed air or inert gases.
Part 3	Safety valves for process fluids.
ISO 4126	Safety valves.

STANDARDS FOR THE INSPECTION AND TESTING OF VALVES

| MSS SP-25 | Standard marking system for valves, fittings, flanges, and unions. |

MSS SP-53	Quality standard for steel castings and forgings for valves, flanges and fittings, and other piping components—magnetic particle examination method.
MSS SP-54	Quality standard for steel castings for valves, flanges and fittings, and other piping components—radiographic examination method.
MSS SP-55	Quality standard for steel castings for valves, flanges and fittings, and other piping components—visual method for evaluation of surface irregularities.
MSS SP-61	Pressure testing of steel valves.
MSS SP-82	Valve-pressure testing methods.
MSS SP-93	Quality standard for steel castings for valves, flanges and fittings, and other piping components—liquid penetrant examination method.
MSS SP-94	Quality standard for steel castings and forgings for valves, flanges and fittings, and other piping components—ultrasonic examination method.
MSS SP-111	Quality standard for evaluation of cast steel surface finishes—visual and tactile method.
MSS SP-121	Qualification testing methods for stem packing for rising stem valves.
API Spec. 6FA	Fire test for valves.
API Spec. 6FC	Fire test for valves with automatic backseats.
API Std. 607	Fire test for soft-seated quarter turn valves.
ASME/API 527	Commercial seat tightness of safety relief valves with metal-to-metal seats.
API Std 598	Valve inspection and test.
BS 6755-1	Testing of valves; Part 1: Specification for production pressure testing requirements.
BS 6755-2	Testing of valves; Part 2: Specification for fire type testing requirements.

MISCELLANEOUS STANDARDS PERTAINING TO VALVES

BS 4371	Fibrous gland packings.

STANDARDS PERTAINING TO RUPTURE DISCS

ASME Code, Section VIII, Division, 1, UG 125 through 136

BS 2915	Bursting discs and bursting-disc devices.
ISO 6718	Bursting discs and bursting-disc devices.
ANSI/NFPA 68	Explosion venting.
VDI 3673	Pressure release of dust explosions.

Appendix D

INTERNATIONAL SYSTEM OF UNITS (SI)

SI UNITS

The international system of units is based upon:

1. Seven base units (Table D-1)
2. Two supplementary units (Table D-2)
3. Derived units.

The derived units may be divided into three groups:

1. Units which are expressed in terms of base and supplementary units (Table D-3)
2. Units which have been given special names and symbols (Table D-4)
3. Units which are expressed in terms of other derived units (Table D-5).

Decimal multiples and sub-multiples may be formed by adding prefixes to the SI units (Table D-6).

SI UNITS CONVERSION FACTORS

Table D-7 gives the conversion factors for Imperial, metric, and SI units.

Table D-1
Base Units of SI

length	meter	m
mass	kilogram	kg
time	second	s
electric current	ampere	A
temperature	kelvin	K
luminous intensity	candela	cd
amount of substance	mole	mol

Table D-2
Supplementary Units of SI

plane angle	radian	rad
solid angle	steradian	sr

Table D-3
Some Derived Units Expressed in Terms of Base and
Supplementary Units

acceleration	meter per second squared	m/s^2
angular acceleration	radian per second squared	rad/s^2
area	square meter	m^2
coefficient of linear expansion	1 per kelvin	$1/k$
density	kilogram per cubic meter	kg/m^3
kinematic viscosity	square meter per second	m^2/s
mass flow rate	kilogram per second	kg/s
molar mass	kilogram per mole	kg/mol
specific volume	cubic meter per kilogram	m^3/kg
velocity	meter per second	m/s
volume	cubic meter	m^3

Table D-4
Some Derived Units Having Special Names

| force | newton
N | $1N$ | $= 1\ kg \cdot m/s^2$ | |
|---|---|---|---|---|
| pressure
stress | pascal
Pa | $1Pa$ | $= 1N/m^2$ | $= 1\ kg\ m \cdot s$ |
| energy
work
quantity of heat
radiant energy | joule
J | $1J$ | $= 1N \cdot m$ | $= 1kg \cdot m^2/s^2$ |
| power
radiant flux | watt
W | $1W$ | $= 1\ J/s$ | $= 1kg \cdot m^2 s^3$ |
| potential difference
electromotive force
electric potential | volt
V | $1V$ | $= 1\ W/A$ | $= 1\ kg \cdot m^2/A \cdot s^3$ |

Table D-5
Some Derived Units Expressed in Terms of Other Derived Units

dynamic viscosity	$Pa \cdot s$	$= kg/m \cdot s$
entropy	J/K	$= kg \cdot m^2 s^2 \cdot K$
heat capacity	J/K	$= kg \cdot m^2/s^2 \cdot K$
heat flux density	W/m^2	$= kg/s^3$
molar energy	J/mol	$= kg \cdot m^2/s^2 \cdot mol$
molar entropy	$J/mol \cdot K$	$= kg \cdot m^2/ \cdot K\ mol$
molar heat capacity	$J/mol \cdot K$	$= kg \cdot m^2/s^2 \cdot Kmol$
moment of force	$N \cdot m$	$= kg \cdot m^2/s^2$
radiant intensity	W/sr	$= kgm^2/s^3 sr$
specific energy	J/kg	$= m^2/s^2$
specific entropy	$J/kg \cdot K$	$= m^2 s^2 \cdot K$
specific heat capacity	$J/kg \cdot K$	$= m^2/s^2 \cdot K$
specific latent heat	J/kg	$= m^2/s^2$
surface tension	N/m	$= kg/s^2$
torque	$N \cdot m$	$= kg \cdot m^2/s^2$

Table D-6
Some SI Prefixes

10^9	giga	G
10^6	mega	M
10^3	kilo	k
10^2	hecto	h
10	deka	da
10^1	deci	d
10^2	centi	c
10^3	milli	m
10^6	micro	μ
10^9	nano	n

Table D-7
Imperial, Metric, and SI Units Conversion Factors Length

mm	cm	in.	ft	yd	m	km	mile
*1	*0.1	0.0393701	3.2808×10^{-3}	1.0936×10^{-3}	$*10^{-3}$		
*10	*1	0.393701	0.032808	0.010936	*0.01		
* 25.4	*2.54	*1	0.083333	0.027778	*0.0254		
*304.8	*30.48	*12	*1	0.333333	*0.3048	$*3.048 \times 10^{-4}$	1.894×10^{-4}
*914.4	*91.44	*36	*3	*1	*0.9144	$*9.144 \times 10^{-4}$	5.682×10^{-4}
*1000	*100	39.3701	3.28084	1.09361	*1	$*10^{-3}$	6.214×10^{-4}
$*10^6$	*100000	39370.1	3280.84	1093.61	*1000	*1	0.621371
1.60934×10^6	160934	*63360	*5280	*1760	1609.34	1.60934	*1

1 thou = *0.0254 mm.
1 Å (ångström) = 10^{10} m.
1 UK nautical mile = 6080 ft = 1853.2 m.
1 international nautical mile = 6076.1 ft = *1852 m.
1 μm (micron) = 10^6 m = 39.37×10^6 in.
Note: starred numbers are exact conversions.

Area

mm²	cm²	in²	ft²	yd²	m²	acre	(hectare) hg	km²	mile²
1	0.01	1.550×10^{-3}	1.076×10^{-5}	1.196×10^{-6}	10^{-6}				
100	1	0.1550	1.076×10^{-3}	1.196×10^{-4}	10^{-4}				
645.16	6.4516	1	6.944×10^{-3}	7.716×10^{-4}	6.452×10^{-4}				
92903	929	144	1	0.1111	0.09290	2.30×10^{-5}	9.29×10^{-6}	9.29×10^{-8}	3.587×10^{-8}
836127	8361	1296	9	1	0.8361	2.066×10^{-4}	8.361×10^{-5}	8.361×10^{-7}	3.228×10^{-7}
10^6	10000	1550	10.764	1.196	1	2.471×10^{-4}	10^{-4}	10^{-6}	3.861×10^{-7}
			43560	4840	4047	1	0.4047	0.4047×10^{-3}	1.562×10^{-3}
			107639	11960	10000	2.471	1	0.01	3.861×10^{-3}
			1.0764×10^7	1.196×10^6	10^6	247.1	100	1	0.386
			2.7878×10^7	3.0976×10^6	2.590×10^6	640	259.0	2.590	1

1 acre = 100 m².

Volume

mm³	*ml	in³	*l	US gal	UK gal	ft³	yd³	m³
1	10^{-3}	6.1024×10^{-5}	10^{-6}	2.642×10^{-7}	2.200×10^{-7}	3.531×10^{-8}	1.308×10^{-9}	10^{-9}
10^3	1	0.061026	10^{-3}	2.642×10^{-4}	2.200×10^{-4}	3.532×10^{-5}	1.308×10^{-6}	10^{-6}
16387	16.39	1	0.01639	4.329×10^{-3}	3.605×10^{-3}	5.787×10^{-4}	2.143×10^{-5}	1.639×10^{-5}
10^6	10^3	61.026	1	0.2642	0.2200	0.03532	1.308×10^{-3}	10^{-3}
3.785×10^6	3785	231.0	3.785	1	0.8327	0.1337	4.951×10^{-3}	3.785×10^{-3}
4.546×10^6	4546	277.4	4.546	1.201	1	0.1605	5.946×10^{-3}	4.546×10^{-3}
2.832×10^7	2.832×10^4	1728	28.32	7.4805	6.229	1	0.03704	0.02832
7.6456×10^8	7.6453×10^5	46656	764.53	202.0	168.2	27	1	0.76456
10^9	10^6	61024	10^3	264.2	220.0	35.31	1.308	1

*1l = 1.000028 dm³ and 1 ml = 1.000028 cm³ according to the 1901 definition of the liter.
1 US barrel = 42 US gal = 34.97 UK gal.
1 fluid oz = 28.41 ml.
1 UK pint = 568.2 ml.
1 liter = 1.760 UK pints.

Volume Rate of Flow (Volume/Time)

litres/h	ml/s	m³/d	l/min	m³/h	Ft³/min	l/s	ft³/s	m³/s
1	0.2778	0.024	0.01667	1×10^{-3}	5.886×10^{-4}	2.778×10^{-4}	9.810×10^{-6}	2.778×10^{-7}
3.6	1	0.08640	0.0600	3.6×10^{-3}	2.119×10^{-3}	1×10^{-3}	3.532×10^{-5}	1×10^{-6}
4.546	1.263	0.1091	0.07577	4.546×10^{-3}	2.676×10^{-3}	1.263×10^{-3}	4.460×10^{-5}	1.263×10^{-6}
41.67	11.57	1	0.6944	0.04167	0.02452	0.01157	4.087×10^{-4}	1.157×10^{-5}
60	16.67	1.44	1	0.0600	0.03531	0.01667	5.886×10^{-4}	1.667×10^{-5}
272.8	75.77	6.547	4.546	0.2728	0.1605	0.07577	2.676×10^{-3}	7.577×10^{-5}
1000	277.8	24	16.67	1	0.5886	0.2778	9.810×10^{-3}	2.778×10^{-4}
1699	471.9	40.78	28.31	1.699	1	0.4719	0.01667	4.719×10^{-4}
3600	1000	86.40	60	3.6	2.119	1	0.03531	1×10^{-3}
1.019×10^{5}	2.832×10^{4}	2446	1699	101.9	60	28.32	1	0.02832
1.854×10^{5}	5.261×10^{4}	4546	3157	189.4	111.5	52.61	1.858	0.05261
3.6×10^{6}	1×10^{6}	5.64×10^{4}	6×10^{4}	3600	2119	1000	35.31	1

Mass

g	oz	lb	kg	cwt	US ton (short ton)	t (tonne)	UK ton
1	0.035274	2.2046×10^{-3}	10^{-3}				
28.3495	1	0.0625	0.028350				
453.592	16	1	0.453592	8.9286×10^{-3}	5.00×10^{-4}	4.5359×10^{-4}	4.4643×10^{-4}
10^3	35.2740	2.20462	1	0.019684	1.1023×10^{-3}	10^{-3}	9.8421×10^{-4}
50802.3	1792	112	50.8023	1	0.056	0.05080	0.05
907185	32000	2000	907.185	17.8571	1	0.907185	0.892857
10^6	35273.9	2204.62	1000	19.6841	1.10231	1	0.984207
1.01605×10^6	35840	2240	1016.05	20	1.12	1.01605	1

1 quintal = 100 kg.

Mass Rate of Flow (Mass/Time)

lb/h	kg/h	g/s	lb/min	lb/s	kg/s
0.2516	0.1142	0.03171	4.194×10^{-3}	6.990×10^{-5}	3.171×10^{-5}
0.2557	0.1160	0.03222	4.262×10^{-3}	7.103×10^{-5}	3.221×10^{-5}
1	0.4536	0.1260	0.01667	2.778×10^{-4}	1.260×10^{-4}
2.205	1	0.2778	0.03674	6.124×10^{-4}	2.778×10^{-4}
7.937	3.6	1	0.1323	2.205×10^{-3}	1×10^{-3}
60	27.216	7.560	1	1.667×10^{-2}	7.56×10^{-3}
91.86	41.67	11.57	1.531	0.02551	0.01157
93.33	42.34	11.76	1.556	0.02593	0.01176
2205	1000	277.8	36.74	0.6124	0.2778
2240	1016	282.2	37.33	0.6222	0.2822
3600	1633	453.6	60	1	0.4536
7937	3600	1000	132.3	2.205	1

Density (Mass/Volume)

kg/m^3	lb/ft^3	lb/in^3	g/cm^3
1	0.062428	3.8046×10^{-5}	10^{-3}
16.0185	1	5.7870×10^{-4}	0.0160185
99.776	6.22884	3.6046×10^{-3}	0.099776
1000	62.4280	0.036127	1
1328.94	82.9630	0.048011	1.32894
27679.9	1728	1	27.6799

*1 g/cm^3 = 1 kg/dm^3 = 1 t/m^3 = 1.000028 g/ml or
1.000028 kg/liter (based on the 1901 definition of the liter).

Velocity

mm/s	ft/min	cm/s	km/h	ft/s	mile/h	m/s	km/s
*1	0.19685	*0.1	$*3.6 \times 10^{-3}$	3.281×10^{-3}	2.237×10^{-3}	$*10^{-3}$	$*10^{-6}$
*5.08	*1	*0.508	0.018288	0.016667	0.01136	$*5.08 \times 10^{-3}$	$*5.08 \times 10^{-6}$
*10	1.9685	*1	*0.036	0.032808	0.022369	*0.01	$*10^{-5}$
277.778	54.6806	27.7778	*1	0.911344	0.621371	0.277778	2.778×10^{-4}
*304.8	*60	*30.48	*1.09728	*1	0.681818	*0.3048	$*3.048 \times 10^{-4}$
*447.04	*88	*44.704	*1.609344	1.46667	*1	*0.44704	$*4.470 \times 10^{-4}$
*1000	196.850	*100	*3.6	3.28084	2.23694	*1	$*10^{-3}$
$*10^6$	196850	*100000	*3600	3280.84	2236.94	$*10^3$	*1

1 UK knot = 1.853 km/h.
*1 international knot (Kn) = *1.852 km/h.*
Note: starred numbers are exact conversions.

Second Moment of Area

mm^4	cm^4	in^4	ft^4	m^4
1	10^{-4}	2.4025×10^{-6}	1.159×10^{-10}	10^{-12}
10000	1	0.024025	1.159×10^{-6}	10^{-8}
416231	41.623	1	4.8225×10^{-5}	4.1623×10^{-7}
8.631×10^9	863097	20736	1	8.6310×10^{-3}
10^{12}	10^8	2.4025×10^6	115.86	1

Force

pdl	N	lbf	*kgf	kN
1	0.1383	0.0311	0.0141	1.383×10^{-4}
7.233	1	0.2248	0.1020	10^{-3}
32.174	4.448	1	0.4536	4.448×10^{-3}
70.93	9.807	2.2046	1	9.807×10^{-3}
7233	1000	224.8	102.0	1
72070	9964	2240	1016	9.964

* *The kgf is sometimes known as the kilopond (kp).*

Moment of Force (Torque)

pdl ft	lbf in	Nm	lbf ft	kgf m
1	0.3730	0.04214	0.03108	4.297×10^{-3}
2.681	1	0.1130	0.08333	0.01152
23.73	8.851	1	0.7376	0.1020
32.17	12	1.356	1	0.1383
232.7	86.80	9.807	7.233	1
6006	2240	253.1	186.7	25.81
72070	26880	3037	2240	309.7

One Nm $= 10^{-7}$ dyn cm.

Stress

dyn/cm²	N/m²	pdl/ft²	lbf/ft²	kN/m²	lbf/in²	kgf/cm²	*MN/m²	kgf/mm²	h bar
1	0.100	0.06720	2.089×10^{-3}	1×10^{-4}	1.450×10^{-5}	1.020×10^{-6}	1×10^{-7}	1.020×10^{-8}	1×10^{-8}
10	1	0.6720	0.02089	1×10^{-3}	1.450×10^{-4}	1.020×10^{-5}	1×10^{-6}	1.020×10^{-7}	1×10^{-7}
14.88	1.488	1	0.03108	1.488×10^{-3}	2.158×10^{-4}	1.518×10^{-5}	1.488×10^{-6}	1.518×10^{-7}	1.488×10^{-7}
478.8	47.88	32.17	1	0.04788	6.944×10^{-3}	4.882×10^{-4}	4.788×10^{-5}	4.882×10^{-6}	4.788×10^{-6}
1×10^4	1000	672.0	20.89	1	0.14450	0.01020	1×10^{-3}	1.020×10^{-4}	1×10^{-4}
6.895×10^4	6895	4633	144	6.895	1	0.07031	6.895×10^{-3}	7.031×10^{-4}	6.895×10^{-4}
9.807×10^5	9.807×10^4	6.590×10^4	2048	98.07	14.22	1	0.09807	0.01000	9.807×10^{-3}
1×10^7	1×10^6	6.720×10^5	2.089×10^4	1000	145.0	10.20	1	0.1020	0.1000
9.807×10^7	9.807×10^6	6.590×10^6	2.048×10^5	9807	1422	100	9.807	1	0.9807
1×10^8	1×10^7	6.720×10^6	2.089×10^5	10000	1450	102.0	10	1.020	1
1.544×10^8	1.544×10^8	1.038×10^7	3.226×10^5	1.544×10^4	2240	157.5	15.44	1.575	1.544

*$1\ MN/m^2 = 1\ N/mm^2$.

Pressure

dyn/cm²	N/m²	lbf/ft²	m bar	mm hg	in H₂O	kN/m²	in Hg	lbf/in²	*kgf/cm²	bar	atm
1	0.1000	2.089×10^{-3}	1×10^{-3}	7.501×10^{-4}	4.015×10^{-4}	1×10^{-4}	2.953×10^{-5}	1.450×10^{-5}	1.020×10^{-6}	1×10^{-6}	9.869×10^{-7}
10	1	0.0289	0.0100	7.501×10^{-3}	4.015×10^{-3}	1×10^{-3}	2.953×10^{-4}	1.450×10^{-4}	1.020×10^{-5}	1×10^{-5}	9.869×10^{-6}
478.8	47.88	1	0.4788	0.3591	0.1922	0.04788	0.01414	6.944×10^{-3}	4.882×10^{-4}	4.788×10^{-4}	4.726×10^{-4}
1000	100	2.089	1	0.7501	0.4015	0.1000	0.02953	0.01450	1.020×10^{-3}	1×10^{-3}	9.869×10^{-4}
1333	133.3	2.785	1.333	1	0.5352	0.1333	0.03937	0.01934	1.360×10^{-3}	1.333×10^{-3}	1.316×10^{-3}
2491	249.1	5.202	2.491	1.868	1	0.2491	0.07356	0.03613	2.540×10^{-3}	2.491×10^{-3}	2.458×10^{-3}
1×10^{4}	1000	20.89	10	7.501	4.015	1	0.2953	0.1450	0.01020	0.0100	9.869×10^{-3}
3.386×10^{4}	3386	70.73	33.86	25.40	13.60	3.386	1	0.4912	0.03453	0.03386	0.03342
6.895×10^{4}	6895	144	68.95	51.71	27.68	6.895	2.036	1	0.07031	0.06895	0.06805
9.807×10^{5}	9.807×10^{4}	2048	980.7	735.6	393.7	98.07	28.96	14.22	1	0.9807	0.9678
1×10^{6}	1×10^{5}	2089	1000	750.1	401.5	100	29.53	14.50	1.020	1	0.9869
1.013×10^{6}	1.013×10^{5}	2116	1013	760.0	406.8	101.3	29.92	14.70	1.033	1.013	1

*1 kgf/cm² = 1 kp/cm² = 1 technical atmosphere.
1 torr = 1 mm Hg (to within 1 part in 7 million).
1 N/m² is sometimes called a pascal.

Energy, Work, Heat

J	ft lbf	*cal	kgf m	kJ	Btu	Chu	*kcal	MJ	hp h	KW h	therm
1	0.7376	0.2388	0.1020	10^{-3}	9.478×10^{-4}	5.266×10^{-4}	2.388×10^{-4}	10^{-6}	3.725×10^{-7}	2.778×10^{-7}	9.478×10^{-9}
1.3558	1	0.3238	0.1383	1.356×10^{-3}	1.285×10^{-3}	7.139×10^{-4}	3.238×10^{-4}	1.356×10^{-6}	5.051×10^{-7}	3.766×10^{-7}	1.285×10^{-8}
4.1868	3.0880	1	0.4270	4.187×10^{-3}	3.968×10^{-3}	2.205×10^{-3}	10^{-3}	4.187×10^{-6}	1.560×10^{-6}	1.163×10^{-6}	3.968×10^{-8}
9.8066	7.2330	2.3420	1	9.807×10^{-3}	9.294×10^{-3}	5.163×10^{-3}	2.342×10^{-3}	9.807×10^{-6}	3.653×10^{-6}	2.724×10^{-6}	9.294×10^{-8}
1000	737.56	238.85	101.97	1	0.9478	0.5266	0.2388	10^{-3}	3.725×10^{-4}	2.778×10^{-4}	9.478×10^{-6}
1055.1	778.17	252.00	107.59	1.0551	1	0.5556	0.2520	1.055×10^{-3}	3.930×10^{-4}	2.931×10^{-4}	10^{-5}
1899.1	1400.7	453.59	193.71	1.8991	1.800	1	0.4536	1.899×10^{-3}	7.074×10^{-4}	5.275×10^{-4}	1.800×10^{-5}
4186.8	3088.0	1000	427.04	4.1868	3.9683	2.2046	1	4.187×10^{-3}	1.560×10^{-3}	1.163×10^{-3}	3.968×10^{-5}
10^{6}	737.562	238.846	101.972	1000	947.82	526.56	238.85	1	0.3725	0.2778	9.478×10^{-3}
2.6845×10^{6}	1.9800×10^{6}	641.186	273.745	2684.5	2544.4	1413.6	641.19	2.6845	1	0.7457	0.02544
3.6000×10^{6}	2.65522×10^{6}	859.845	367.098	3600	3412.1	1895.6	859.84	3.600	1.3410	1	0.03412
1.0551×10^{8}	7.78817×10^{7}	2.5200×10^{7}	1.0759×10^{7}	1.0551×10^{5}	100.000	55.556	25200	105.51	39.301	29.307	1

* cal is the International Table calorie.
1 hp h = 1.014 hp h (metric) 745.7 Wh 2.685 MJ.
1 thermie = 1.163 kWh 4.186 MJ 999.7 kcal.
1 ft pdl = 0.04214 J.
1 erg = 10^{-7} J.

Power, Heat Flow Rate

Btu/h	Chu/h	W	kcal/h	ft lbf/s	kgf m/s	metric hp	hp	kW	cal/h	MW
1	0.5556	0.2931	0.2520	0.2162	0.02988	3.985×10^{-4}	3.930×10^{-4}	2.931×10^{-4}	2.520×10^{-4}	2.93×10^{-7}
1.800	1	0.5275	0.4536	0.3892	0.05379	7.172×10^{-4}	7.07×10^{-4}	5.275×10^{-4}	4.536×10^{-4}	5.28×10^{-7}
3.4121	1.8956	1	0.8598	0.7376	0.1020	1.360×10^{-3}	1.341×10^{-3}	10^{-3}	8.598×10^{-4}	10^{-6}
3.9683	2.2046	1.163	1	0.8578	0.1186	1.581×10^{-3}	1.560×10^{-3}	1.163×10^{-3}	10^{-3}	1.16×10^{-6}
4.626	2.5701	1.3558	1.1658	1	0.1383	1.843×10^{-3}	1.818×10^{-3}	1.356×10^{-3}	1.166×10^{-3}	1.36×10^{-6}
33.46	18.59	9.807	8.432	7.233	1	0.01333	0.01315	9.807×10^{-3}	8.432×10^{-3}	9.81×10^{-6}
2510	1394	735.50	632.4	542.5	75	1	0.9863	0.7355	0.6324	7.355×10^{-4}
2544	1414	745.70	641.19	550	76.040	1.0139	1	0.7457	0.6412	7.457×10^{-4}
3412.1	1896	1000	859.8	737.6	102.0	1.360	1.3410	1	0.8598	10^{-3}
3968.3	2204.6	1163	1000	857.8	118.6	1.581	1.5596	1.163	1	1.163×10^{-3}
3.4121×10^{6}	1.896×10^{6}	10^{6}	8.598×10^{5}	7.376×10^{5}	1.0197×10^{5}	1360	1341	1000	859.8	1

$1 W - 1 J/S$.
$1\ cal/s = 3.6\ kcal/h$.
1 ton of refrigeration $= 3517\ W = 12000\ Btu/h$.
$1\ erg/s = 10^{7}\ W$.

Dynamic Viscosity

μ Ns/m²	kg/mh	lb/ft h	cP	P	Ns/m²	pdl s/ft²	kgf s/m²	lbf s/ft²	lbf h/ft²
1	3.6×10^{-3}	2.419×10^{-3}	10^{-3}	10^{-5}	10^{-6}	6.720×10^{-7}	1.020×10^{-7}	2.09×10^{-8}	5.80×10^{-12}
277.8	1	0.6720	0.2778	2.778×10^{-3}	2.778×10^{-4}	1.866×10^{-4}	2.832×10^{-5}	5.80×10^{-6}	1.61×10^{-9}
413.4	1.488	1	0.4134	4.134×10^{-3}	4.134×10^{-4}	2.776×10^{-4}	4.215×10^{-5}	8.63×10^{-6}	2.40×10^{-9}
1000	3.6	2.419	1	0.010	10^{-3}	6.720×10^{-4}	1.020×10^{-4}	2.09×10^{-5}	5.80×10^{-9}
10^{5}	360	241.9	100	1	0.100	0.0672	0.01020	2.089×10^{-3}	5.80×10^{-7}
10^{6}	3600	2419	1000	10	1	0.6720	0.1020	0.02089	5.80×10^{-6}
1.138×10^{6}	5358	3600	1488	14.88	1.488	1	0.15175	0.03108	8.63×10^{-6}
9.807×10^{6}	3.530×10^{4}	2.372×10^{4}	9807	98.07	9.807	6.590	1	0.2048	5.69×10^{-5}
4.788×10^{7}	1.724×10^{5}	1.158×10^{5}	47880	478.8	47.88	32.17	4.882	1	2.778×10^{-4}
1.724×10^{11}	6.205×10^{8}	4.170×10^{8}	1.724×10^{8}	1.724×10^{6}	1.724×10^{5}	1.158×10^{5}	1.758×10^{4}	3600	1

1 cp = 1 mN s/m² = 1 g/ms.
1 P = 1 g/cm s = 1 dyn s/cm².
1 N s/m² = 1 kg/m s.
1 pdl s/ft² = 1 lb/ft s.
1 lbf s/ft² = 1 slug/ft s.

Kinematic Viscosity

in²/h	cSt (mm²/s)	ft²/h	St (cm²/s)	m²/h	in²/s	ft²/s	m²/s
1	0.1792	6.944×10^{-3}	1.792×10^{-3}	6.452×10^{-4}	2.778×10^{-4}	1.93×10^{-6}	1.79×10^{-7}
5.5800	1	0.03875	0.010	3.60×10^{-3}	1.550×10^{-3}	1.076×10^{-5}	10^{-6}
144	25.81	1	0.2581	0.0929	0.04	2.778×10^{-4}	0.258×10^{-4}
558.0	100	3.8750	1	0.36	0.1550	1.076×10^{-3}	10^{-4}
1550	277.8	10.76	2.778	1	0.4306	2.990×10^{-3}	2.778×10^{-4}
3600	645.2	25	6.452	2.323	1	6.944×10^{-3}	6.452×10^{-4}
518400	92903	3600	929.0	334.5	144	1	0.0929
5.580×10^{6}	10^{6}	3.875×10^{4}	10000	3600	1550	10.76	1

Density of Heat Flow Rate (Heat/Area × Time)

W/m²	kcal/m²h	Btu/ft²h	Chu/ft²h	kcal/ft²h	KW/m²
1	0.8598	0.3170	0.1761	0.07988	10^{-3}
1.163	1	0.3687	0.2049	0.09290	1.163×10^{-3}
3.155	2.712	1	0.5556	0.2520	3.155×10^{-3}
5.678	4.882	1.800	1	0.4536	5.678×10^{-3}
12.52	10.76	3.968	2.205	1	0.01252
1000	859.8	317.0	176.1	79.88	1

Heat-Transfer Coefficient
(Thermal Conductance; Heat/Area × Time × Degree Temperature)

W/m²°C	kcal/m²h°C	Btu/ft²h°F	kcal/ft²°C	kW/m²°C	Btu/ft²s°F	cal/cm²s°C
1	0.8598	0.1761	0.07988	10^{-3}	4.892×10^{-5}	2.388×10^{-5}
1.163	1	0.2048	0.09290	1.163×10^{-3}	5.689×10^{-5}	2.778×10^{-5}
5.678	4.882	1	0.4536	5.678×10^{-3}	2.778×10^{-4}	1.356×10^{-4}
12.52	10.76	2.205	1	0.01252	6.124×10^{-4}	2.990×10^{-4}
1000	859.8	176.1	79.88	1	0.04892	0.02388
20442	17577	3600	1633	20.44	1	0.4882
41868	36000	7373	3344	41.87	2.048	1

$1\ Btu/ft^2\ h°F = 1\ Chu/ft^2\ h°C.$
$1\ W/m^2°C = 10^{-4}\ W/cm^2°C.$

Thermal Conductivity
(Heat × Length/Area × Time × Degree Temperature)

Btu in/ft²h°F	kcal in/ft²h°C	W/m°C	kcal/m h°C	Btu /ft h°F	cal/cm s°C
1	0.4536	0.1442	0.1240	0.0833	3.445×10^4
2.2046	1	0.3180	0.2734	0.1837	7.594×10^4
6.933	3.146	1	0.8598	0.5778	2.388×10^3
8.064	3.658	1.163	1	0.6720	2.778×10^3
12	5.443	1.731	1.488	1	4.134×10^3
2903	1317	418.7	360	241.9	1

1 Btu in/ft² h°F = 1 Chu in/ft² h°C.
1 Btu/ft h°F = 1 Btu ft/ft² h°F = 1 Chu/ft h°C.
1 W/m°C = 10^2 W/cm°C = 1 kW mm/m² °C.

Specific Heat Capacity (Heat/Mass × Degree Temperature)

ft lbf/lb°F	kgf m/kg°C	kJ/kg°C	*Btu/lb°F	kcal/kg°C
1	0.5486	5.380×10^{-3}	1.285×10^{-3}	1.285×10^{-3}
1.823	1	9.807×10^{-3}	2.342×10^{-3}	2.342×10^{-3}
185.9	101.97	1	0.2388	0.2388
778.2	426.9	4.1868	1	1
778.2	426.9	4.1868	1	1

* *1 Btu/lb° F=1 Chu /lb° C.*

Specific Energy
(Heat/Mass; e.g., Calorific Value, Mass Basis, Specific Latent Heat)

ft lbf/lb	kgf m/kg	*kJ/kg	Btu/lb	kcal/kg	MJ/kg
1	0.3048	2.989×10^{-3}	1.285×10^{-3}	7.139×10^{-4}	2.989×10^{-6}
3.281	1	9.807×10^{-3}	4.216×10^{-3}	2.342×10^{-3}	9.807×10^{-6}
334.55	101.97	1	0.4299	0.2388	10^{-3}
778.2	237.19	2.326	1	0.556	2.236×10^{-3}
1400.7	426.9	4.187	1.8	1	4.187×10^{-3}
334553	101972	1000	429.9	238.8	1

**p1 J/g=1 kJ/kg.*
1 kcal/kg=1 Chu /lb.

Calorific Value, Volume Basis (Heat/Volume)

J/m³	kJ/m³	kcal/m³	Btu/ft³	Chu/ft³	*MJ/m³
1	1×10^{-3}	2.388×10^{-4}	2.684×10^{-5}	1.491×10^{-5}	1×10^{-6}
1000	1	0.2388	0.02684	0.01491	1×10^{-3}
4.187×10^3	4.187	1	0.1124	0.06243	4.187×10^{-3}
3.726×10^4	37.26	8.899	1	0.5556	0.03726
6.707×10^4	67.07	16.02	1.800	1	0.06707
1×10^6	1000	238.8	26.84	14.91	1

1 therm (10^5 Btu) UK gal = 2320 8 MJ m³.
1 thermie/liter = 4185 MJ m³.
*MJ/m³ = J/cm³.

Appendix E

VALVE GLOSSARY

There are a variety of terms used in the valve industry and the purpose of this appendix is to assist the reader. This valve glossary is a list of definitions, describing the types of valves, their functions, individual components, their materials of construction, and their design characteristics.

Accessory. A device attached to the actuator which provides an additional function, for example manual operation, positioner, etc.

Actual pressure drop. The difference between the inlet pressure and the outlet pressure of a valve.

Actuator. A device used to open/close or control the valve. Key types include electrical, hydraulic, and pneumatic. Movement may be quarter-turn or multi-turn. Actuators may be used when

* valves are remotely located (e.g., on pipelines)
* valves are located in hazardous areas
* manual operation would be time-consuming (e.g., with larger valves)

Actuator stem. A rod used in linear designed valves connecting the actuator with the stem of the valve.

Actuator stem force. The amount of force that is required to move the actuator stem to either open or close the valve.

Air filter. An accessory added to an actuator to prevent oil, dirt, or water in the air supply from entering the pneumatic actuator.

Air valve. Valve that is used to control the flow of air. Flows are normally small, so solenoid valves are suited.

Angle valve. A valve body where the inlet and outlet ports are at 90° to each other.

Arithmetic average roughness height. The measurement of the smoothness of a surface. In the case of valves refers to the flange faces and is usually given in microns or AARH: the smaller the value the smoother the surface is.

Back flow. When the normal process is reversed.

Back pressure. The pressure exerted on the downstream side of a valve seat.

Back seats. In linear valves, the area of the stem that enters the valve bonnet is sealed to prevent process fluid from entering the packing box and to prevent deterioration of the sealing materials.

Ball valve. A quarter-turn valve with a spherical closing element held between two scats. Characteristics include quick opening and good shut-off. Ball valves are widely used as on/off valves in the chemical process and other industries. Special designs (with V notches or fingers) are available for throttling applications. Larger valves with heavier balls (e.g., on pipelines) may use trunnions to help support the ball and prevent damage to soft internals. Designs are typically one, two, or three piece.

Bellows. A sealing device that prevents line media leaking between the stem and the body.

Blowdown. The discharge of process fluid to reduce the pressure in a piping system. This is usually done through a pressure relief device.

Body. The main pressure-containing component; contains the closure device.

Bonnet. The pressure-containing component that contains the packing box and the stem. It can be screwed, flanged, or welded to the body of the valve.

Breakout torque. The torque required to open or unseat a rotary valve.

Brinell hardness number. A number from 111 to 745 that indicates the relative hardness of a material. As the number increases the harder the material is said to become.

Bubble-tight. When there is no measurable seat leakage over a certain period of time during test conditions.

Butterfly valve. A quarter-turn valve, which has a circular disc as its closing element. The standard design has the valve stem running through the center of the disc, giving a symmetrical appearance. Later more complex designs offset the stem, so that the disc "cams" into the valve seat. Advantages include less wear and tear on the disc and seats, and tighter shut-off capabilities. Many design types are available including inexpensive Teflon® or resilient seats for use in water (treatment) plants, etc. More expensive metal seats can be used where high temperatures or aggressive chemicals are encountered. So-called "high-performance" butterfly valves offer zero leakage designs and have been applied in both the chemicals and hydrocarbon processing sectors.

Butt-weld end connection. A special end connection that is beveled to allow welding to a similarly beveled piece of pipe to allow a full penetration weld to be made.

Bypass valve. A valve smaller in diameter that is fitted in parallel to a larger main valve. Bypass valves are used to reduce the differential pressure across the main valve before this main valve is opened (as otherwise this larger, more expensive valve, may suffer damage to internal components). In some services they are used to warm up the downstream side of the valve, before opening the larger valve.

Certified dimensional drawing. A drawing that guarantees the overall dimensions of the valve that are required for installation. Sometimes called the general arrangement or GA.

Certified material test report. Information on the component that covers its chemical composition and its mechanical properties.

Chainwheel. A handwheel design that has sprockets that allow a chainwheel to be wrapped around a semicircle section of the handwheel and used as a pulley to turn the stem. This is installed on valves that are installed at an elevated position where it is not possible to erect a platform or add a ladder.

Check valve. A valve that is designed to allow the fluid to flow in a given direction but closes to prevent back flow. Types include swing check, tilting-disc check, and wafer check, non slam (piston type). Check valves (also called non-return valves) are usually self-acting.

Class. The class is used to describe the pressure rating of the piping system. For example Class 150 lb, 300 lb, 600 lb, 900 lb, or API 3000, API 5000. This relates to the maximum allowable design pressure that a flange of certain dimensions and made of a certain material can be used with in a piping system.

Concentric butterfly valve. A butterfly valve with the disc installed in the center of the valve.

Control valve. A valve that regulates the flow or pressure of a fluid. Control valves normally respond to signals generated by independent devices such as flow meters or temperature gauges. Control valves are normally fitted with actuators and positioners. Pneumatically actuated globe valves are widely used for control purposes in many industries, although quarter-turn types such as (modified) ball and butterfly valves may also be used.

Corrosion. The deterioration of a metal that is caused by a chemical reaction. This is sometimes called "weight loss."

Cryogenic valves. Valves suited for use at temperatures below −45 degrees Celsius. A cryogenic valve should have a cold box as an integral part of the body to allow a vapor barrier to form between the packing box and the liquified gas.

C_V. The C_V of a valve is defined as 1 U.S. gallon of 60°F water during 1 minute with a 1 psi pressure drop. Also known as the valve coefficient or the flow coefficient.

Cylinder. A pressure-containing component and the part of an actuator that houses a piston that will be powered either pneumatically or hydraulically.

Design pressure. The pressure used during the design of a piping system, and defines the criteria for pipe wall thickness, fittings, flanges, valves, bolt torque, and threads.

Design temperature. The temperature used during the design of a piping system, and defines the criteria for pipe wall thickness, fittings, flanges, valves, bolt torque, and threads.

Destructive test. A test during which all or part of a component is destroyed by mechanical or chemical means to discover its properties.

Diaphragm valve. A bi-directional valve that is operated by applying an external force to a flexible element or a diaphragm (typically an elastomer). Diaphragm valves may be used for slurries (where other valve designs might clog) or in hygienic applications.

Differential pressure. The pressure difference between the upstream and the downstream ports of a valve. Also called the delta P.

Direct-acting actuator. A diaphragm actuator that allows the actuator stem to extend.

Disc. The closure component in a butterfly valve (rotary) or a globe valve (linear).

Diverter valve. A valve that can change the direction of the flow of a medium to two or more different directions.

DN. The ISO standard abbreviation for the nominal diameter of the line pipe size in metric units. 4" = 100 DN.

Double-acting positioner. A positioner that has the facility to supply and exhaust air on both sides of the actuator piston or diaphragm at the same time.

Double block and bleed. A valve configuration in which positive shut-off is achieved at both the inlet and outlet sides. A small port is fitted to discharge fluid in the intermediate space. Fitting a gas detector to the port provides assurance of the integrity of the upstream seal. This configuration is often required to isolate high-pressure sections of a system to facilitate safe maintenance, etc.

Double disc check valve. A check valve with two semicircular discs that are hinged together and that fold together when the flow is in the correct direction and swing closed when the flow is reversed. Also known as a split disc check valve.

Downstream. The process stream after it has passed through the valve.

Drop tight. A bubble-tight test that involves a water-under-air test.

Ductility. The characteristic of a metal to deform when placed under force. Ductility is measured by the percentage increase of a stretched test piece, just prior to fracture.

Dye penetrant. A bright red or fluorescent dye that is used to detect surface cracks, pitting, or porosity. It is applied by spray and the excess dye is wiped away to expose surface flaws that can be detected by natural or fluorescent light.

Eccentric butterfly valve. A butterfly valve where the shaft that carries the closure disc is slightly offset and creates an elliptical motion as it leaves the sealing surface. This effect reduces friction and wear to the closure disc.

Elastomer. A polymer that is both flexible and resilient when used as a seal.

Electric actuator. An actuator that uses an electric motor to operate the valve stem.

Electrohydraulic actuator. An actuator that supplies hydraulic power to control the valve, but has an electric power source.

End connection. The part of the valve that joins to the piping system. This could be screwed, socket weld, flanged, butt weld, clamped, soldered.

End to end. The extremities of the valve. One connection to the other end connection.

Erosion. Material weight loss inside a piping system, caused by the process flow. This is not a consideration in process flows that have been adequately filtered and where entrained solids are not present.

Examination. The review of a complete valve or its individual components to confirm that it complies with the user's requirements.

Explosion proof. An assurance that an electrical device can be used in an area that is potentially explosive. This device must be detached from any electrical source that might arc.

Extended bonnet. Used when the medium is at high or low temperatures, to avoid damage to the sealing elements.

Fail closed. An actuator facility such that in the event of power failure the valve will move to the fully closed position.

Fail open. An actuator facility such that in the event of power failure the valve will move to the fully open position.

Fail-safe. An actuator facility such that in the event of power failure the valve will move to a predetermined position, which could be open, closed, or an intermediate position.

Fire resistant. The ability of a valve to withstand a fire and maintain the failure position. Such a valve will be equipped with devices to achieve this status.

Fire-safe. The ability of a valve to minimize the amount of process lost downstream or to the atmosphere after a fire test.

Flashing. Caused when the pressure at the vena contracta falls below the vapor pressure, followed by a pressure recovery that is maintained. This creates vapor bubbles that continue downstream. The liquid/gas mixture increases the velocity of the process stream, which can result in excessive noise.

Flat face. A flange that has no raised face or a ring groove surface. These flanges are generally used in lower piping pressure classes such as ASME 125 lb or 150 lb in cast iron and carbon steel. The mating gasket will be flat and extend to the circumference with holes to accommodate the flange bolting.

Flat gasket. A circular, flat sheet with an inside and outside diameter. An annulus.

Floating ball. A ball valve where the closure ball is not attached to the body of the valve.

Floating seat. A seat ring that is not attached to the valve body and can move to suit the closure element and improve the shut-off.

Float valve. A valve that automatically opens or closes as the level of a liquid changes. The valve is operated mechanically by a float that rests on the top of the liquid.

Fluid. A material that can flow; includes gases, liquids, slurries, pellets, and powders.

Full bore. Term used for example of a ball valve, to indicate that the internal diameter of the valve opening is the same as that of the piping to which it is fitted.

Full-bore valve. Any valve where the closure element has the same inside diameter as the inlet and outlet of the valve. Also called a full-port valve and has a lower pressure drop than a reduced bore.

Full closed. The position of the valve when the closure element is fully seated.

Full lift. When a pressure relief valve is fully open upon overpressurization of the piping system.

Full open. The position of the valve when the closure element is fully open allowing maximum flow through the valve.

Full trim. The area of the valve's seat that can pass the maximum flow for that particular size.

Galling. The damage of two mating parts when microscopic portions impact and make a temporary bond. When effort is made to separate these two surfaces, tearing of the two components can occur. This usually happens when the two materials are the same or possess several very similar mechanical characteristics.

Gasket. A soft or a hard sealing material used in conjunction with flanges.

Gate valve. A multi-turn valve that has a gate-like disc and two seats to close the valve. The gate moves linearly, perpendicular to the direction of flow. This type of valve is normally used in the fully opened or fully closed position; it is not suited to throttling applications. Gate valves provide robust sealing, and are used extensively in the petrochemicals industries. This class of valve also includes knife gate valves, conduit gate valves, and wedge gate valves. Knife gate valves have much thinner gates with a knife-like edge, making them suited to use with floating solids, for example as in the pulp and paper industries. Conduit gate valves have a rectangular disc as the closing element. One half of the disc is solid, to close the valve, the other half has a circular port, which can be used to open the valve. Wedge gate valves have a wedge-shaped gate, which "wedges" between floating seats to close the valve tightly.

Gearboxes. Used to ensure easier operation of larger valves, particularly ball valves.

Gland bushing. Or the packing follower. Located at the top of the packing box, it acts as a barrier, protects the packing from the atmosphere, and transfers a force from the gland flange bolting to the packing.

Gland flange. Part of the valve used to compress and retain the internals in the packing box.

Globe valve. A multi-turn valve with a closing element that moves perpendicularly to the valve body seat and generally seals in a plane parallel to the direction of flow. This type of valve is suited to both throttling and general flow control.

Graphite. A carbon-based gasket or packing material, suitable for ambient and high temperatures.

Hardfacing. The welding of a harder alloy over a softer base metal to create a more resistant surface.

Hardness. A material's ability to resist indentation.

Hardness Rockwell test. Method of testing and registering a material's hardness based on the depth of indentation. The higher the number the greater the hardness. This hardness is identified as HRB or HRC depending on the scale used.

Heat treating. The metal-producing process that involves heating and cooling to predetermined temperatures in a particular order and with specific holding times.

High-performance valves. A valve specifically designed for accurate throttling applications.

Hydraulic actuator. A device fitted to the valve stem which uses hydraulic energy to open, close, or regulate the valve. The hydraulic fluid may, according to the configuration, both open and close the valve, or just open the valve. In the latter case, a spring will typically be fitted inside the actuator to return it (and the valve) to the closed position.

Hydrostatic test. A test using water under pressure to detect any leaks through the body, sealing joints, or closure element. Generally this test pressure is 1.5 times the design pressure at ambient temperature.

Impact. A test that will determine the toughness of a particular material by measuring the force necessary to fracture the test piece.

Inclusion. A foreign object or particles found in a weld, forging, or casting that will have a detrimental effect on the component and cause failure or create a leak path.

Inlet. The port where the fluid enters the valve.

Inspection. The examination of a valve or a component by the end user or an authorized third party inspector. This is to confirm that the valve or component meets the user's requirements.

Integral flange. A flanged connection that is either fabricated or cast to the body of the valve.

Integral seat. A seat that is actually a machined part of the valve body and not one that is inserted into the valve.

Intrinsically safe. An electrical device that is not able to produce sufficient heat to cause ignition in the atmosphere.

Jacketed valve. Valve designed to incorporate a so-called jacket around the valve body. Steam is introduced into the jacket to keep the fluids being controlled at the required temperature.

Leakage. Process fluid that passes through a valve when it is fully closed.

Lever operator. A manual method of operating a valve that comprises a pivot handle.

Lift check valve. A non-return valve that prevents back flow by having a free floating element, either a ball or a poppet. The design incorporates a piston to damp the disc during operation.

Limit stop. A device in an actuator that limits the linear or rotary motion of an actuator; can be adjusted.

Limit switch. An electromechanical accessory that is attached to an actuator and used to identify the position of a valve's closure element.

Linear valve. A valve that has a sliding stem that pushes the throttling element up and down. See multi-turn.

Line blind. A pipeline shut-off device, whereby a flat disc is forced between two flanges. Line blinds are less expensive than valves, but require much more time to operate.

Locking device. A device that can be attached to a valve or an actuator and that will enable it to be locked closed or locked open. Prevents accidental operation as only authorized personnel can operate the valve.

Lug body. A body of a flangeless wafer butterfly valve that requires bolts to pass through the body to flanges on either side of the valve. These holes can be tapped to allow the line to be dismantled without "dropping" the valve. Tapped lugged valves are sometimes called "end of line" valves.

Magnetic particle inspection. Iron filings are spread over the area under examination. On passing an electric current through the examination piece, the filings will collect where there are imperfections.

Manual handwheel. A handwheel to open, close, or position a closing element, which does not require an actuator to make it function.

Manual valve. A valve that is worked by manual operation, such as a handwheel or a lever. These valves are generally used for on-off service.

Maximum allowable operating pressure. The maximum pressure that can be safely held in a piping system, expressed in bar (kilopascal) or psi. Determined by the material of construction, the maximum operating temperature, and the piping class. Also called the *Maximum allowable working pressure.*

Metal seat. A seat design where the fixed mating surface with movable closure component is made of metal. Metal-to-metal seats have greater leakage rates than soft seated valves, but they can be used at higher temperatures and pressures.

Mill test report. Report of the chemical testing and physical testing performed on a base material. This documentation is normally produced by the manufacturer and is often requested by the purchaser to confirm compliance to the specification.

Multiported. Multiported valves include additional inlet/outlet ports, to allow fluids to be directed. The ball and plug valve types are ideally suited to multiport designs.

Multi-turn. Category of valves (such as gate, globe, needle), which require multiple turns of the stem to move the valve from the fully open to the fully closed position. Also known as linear valves. See also *Quarter-turn.*

National pipe thread. A tapered thread that is used for pressure connections for piping.

Needle valve. Multi-turn valve that derives its name from the needle-shaped closing element. The design resembles that of the globe valve. Typically available in smaller sizes, they are often used on secondary systems for on/off applications, sampling, etc.

Non-destructive examination. A test to determine a characteristic of a piece of material or its reliability in use, without causing any damage or destruction to the material.

Non-return valve. A valve that allows the flow of a process fluid in only one direction. It will not allow any flow reversal.

Non-rising stem. A valve where the stem is threaded and the turning of a stationary operator will result in the closure element rising to open and lowering to close.

Normally closed. A valve that is normally closed during operation. In many cases these valves are locked closed by using a mechanical device.

Normally open. A valve that is normally open during operation. In many cases these valves are locked open by using a mechanical device.

On-off valve. Basic operation for a manual valve used to start or stop the flow of a process fluid.

Operating medium. The power supply used to operate an actuator: can be pneumatic, hydraulic, or electric.

Operating pressure. The pressure at which a valve usually operates under normal conditions. This is lower than the design pressure.

Operating temperature. The temperature at which a valve usually operates under normal conditions. This is lower than the design temperature.

Operator. A device, handwheel, lever, or wrench used to open, close, or position the closing element of a valve.

O-ring. An elastomer ring that forms a sealing material for the internals of a valve.

Packing. A soft sealing material that is used to prevent leakage of process fluid from around the stem. It is located in the packing box.

Packing box. A chamber through which the stem passes. This chamber houses the packing material, packing spacers, lantern rings, guides, and other seal accessories necessary to prevent leakage of process fluid.

Parallel gate valve. A gate valve that has a flat disc gate that slides between two parallel free floating seats.

PEEK. The abbreviation for polyether ether ketone. A robust soft seating material.

Penstock valve. A type of simple gate valve, used to contain fluids in open channels. Often found in waste water treatment plants.

Pilot valve. Small valve requiring little power that is used to operate a larger valve. See also *Solenoid valve.*

Pinch valve. A valve in which a flexible hose is pinched between one or two moving external elements to stop the flow. This valve is often used in slurry and mining applications, as its operation is not affected by solid matter in the medium. It is also used with certain gases, as the absence of possible leak paths to the atmosphere ensures good emission control.

Piping and instrument diagram. A schematic that indicates the process system, and includes items of equipment, valves, and associated instrumentation. Not to scale.

Piping schedule. A method of noting the wall thickness of a pipe, for example Sch 40, Sch 80, Sch 160. The larger the number the thicker the wall thickness of the pipe at a given nominal diameter.

Pitting corrosion. Surface corrosion that appears as small holes or cavities. Over time these cavities will increase in size and join to create larger cavities.

Plug. In globe valves the closure element can be a tapered plug that extends into the seat.

Plug valve. This multi-turn valve derives its name from the rotating plug that forms the closing element. The plug may be cylindrical or truncated. In the open position, the fluid flows through a hole in the plug. Lubricated plug valves rely on a sealing compound injected between the plug and the valve body, whilst sleeved plug valves are fitted with a "soft" insert between the plug and the body.

Pneumatic actuator. A device fitted to the valve stem which uses pneumatic energy to open/close or regulate the valve. The compressed air may, according to the configuration, both open and close the valve, or just open the valve. In the latter case, a spring will typically be fitted inside the actuator to return the valve to the closed position.

Polyethylene. A flexible thermoplastic that is used for valve seats.

Polypropylene. A thermoplastic that is not as flexible as polyethylene.

Poppet. A closure element in a check valve that is held in place by a spring.

Porosity. Small air bubbles that were created in the casting when the metal was molten. When the metal has cooled, these trapped bubbles weaken the structure and can cause failure in the component.

Positioner. A device that receives a signal—pneumatic or electric—from a controller and compares it to the actual position of the valve. If the signal is not correct then the positioner sends pressure to, or bleeds pressure from, the valve so that the correct position is achieved.

Positive material identification. A testing process that will identify the material specimen. It is possible to determine the approximate chemical composition.

Pressure drop. The difference between the upstream pressure and the downstream pressure of a valve.

Pressure reducing valve. A self-operating valve used to reduce any excess pressure in a system, for example steam. Also known as a PRV. The valve opens if the internal pressure exceeds that holding the closing element onto the seat.

Process flow diagram. A schematic that outlines the process in a plant, and which will include major in-line instrumentation and equipment. Pipe sizing and utility piping might not be shown.

Proximity switch. A limit switch that indicates the valve position without making mechanical contact. The switch will use a magnetic or an electronic sensor to determine the valve position.

psi. The abbreviation for pounds per square inch.

psia. The abbreviation for pounds per square inch absolute. The psia unit is used when the pressure is expressed without taking into account ambient pressure.

psig. The abbreviation for pounds per square inch gauge. The psig unit is used when the pressure is expressed to standard atmospheric pressure (noted 14.7 psia).

PT. The abbreviation for penetrant test.

PTFE. The abbreviation used for polytetrafluoroethylene.

Quarter-turn. The 90° angle through which a valve's closing element must move from the fully open position to the fully closed position. Examples are ball, plug, and butterfly valves.

Rack and pinion actuator. An actuator used in conjunction with quarter-turn valves. This actuator will supply either a pneumatic or a hydraulic force to move a flat-toothed rack that turns a gear to open and close the closure element.

Radiography. A method of examination that uses X-rays to produce an internal image of a test piece. The radiographic results on a film will reveal porosity, inclusions, and cracks within the material.

Raised face flange. A flange face that has a raised section on the mating surface. This raised section can come with various types of serrated finish. This allows greater loadings to be applied to the gasket and creates a more efficient seal than a flat face flange.

Reduced bore. Indicates that the internal diameter of the valve is lower than that of the piping to which the valve is fitted.

Reduced-port valve. A valve that has a smaller internal bore than those of the inlet and the outlet. A reduced-port (bore) ball valve will have a greater pressure drop than a full-port (bore) ball valve.

Regulating valve. Valve type used to regulate flows to provide a constant pressure output.

Ring type joint. A flanged end connection with a circular groove on the mating face, where a softer metal ring is placed before mating up to a similar flange face and bolting up. The softer ring, usually oval or hexagonal, will deform when the flanges are bolted up and create a tight seal. Ring type joint connections are used on higher-pressure piping systems, ASME 900 lb and above. The abbreviation is RTJ.

Safety valve. A pressure relief valve that is designed to reduce overpressurization in a gas or steam service.

Sampling valve. A valve that is fitted to a reactor or pipeline to allow small samples of a fluid to be withdrawn for further testing. In simple cases a standard gate or needle valve, for example, may be used. The disadvantage is that inappropriate use may result in spillage. As an alternative, valves are available which "trap" a small quantity of fluid in a chamber and only this small amount of fluid is released when the valve is operated.

Screwed bonnet. A valve bonnet with male threads to join a valve body with female threads.

Screwed end connections. End connections that have female national pipe thread (NPT), which mates with male NPT on a pipe.

Seal load. For linear valves, the force that must be generated by an actuator on a stem to overcome the various forces acting on the shaft during the opening, closing, or positioning of the closure element.

Seal weld. The fillet type weld required for socket weld fittings to prevent leakage.

Seat. A circular ring into which the closure element of a globe valve enters. This element is a plug, needle, or disc. The plug/needle/disc enters the circular ring (seat).

Seat pressure differential. The difference between the operating pressure and the set pressure for the system; the set pressure is the higher.

Seating torque. The torque value produced by a rotary actuator to open or close the valve.

Shaft. The rod that connects the closure element and the closure operator (handwheel or actuator).

Shut-off. When the valve is in a closed position and flow ceases.

Shut-off valve. The valve to achieve shut-off.

Single-acting actuator. An actuator in which air is applied to one chamber. This air pressure acts against and pushes a plate.

Sliding gate valve. A gate valve that has a flat rectangular plate as a closure element. Sometimes called a sluice valve and used for large bore irrigation and waterworks systems.

Slurry. A process fluid that contains undissolved solids.

Soft seat plug. An elastomer that is placed within a metal ring at the seating area of a globe valve. This will provide a bubble-tight shut-off.

Solenoid. An accessory to an actuator that acts as a control device. It can regulate the air supply to an actuator for on-off or throttling of the valve.

Solenoid valve. Valve, typically of the needle globe type, that is operated by an electrical solenoid. Such valves are often deployed as pilot valves, that is, fitted to actuators that in turn control larger valves.

Speed of response. The speed provided by an actuator to operate a valve. Sometimes called the stroking speed.

Spiral wound gasket. A gasket that contains hard and soft elements to create a seal. A stainless-steel strip is coiled to create a circular disc with small spaces that are then filled with graphite or another soft non-metallic material. A spiral wound gasket is held between two flanges and bolted up.

Split body. Usually refers to a ball valve that comprises more than one piece and houses the closure element.

Spring. In diaphragm actuators, this is the component that applies the force to act against the piston in the chamber. It provides the force necessary to move the closure element to the correct failure position.

Spring rate. The amount of force generated by a spring when it is compressed to a certain measurement.

Spring return. See *Pneumatic actuator.*

Stroke. The travel required by a valve, either linear type or rotary type, generally from fully open to fully closed.

Sub sea valve. A valve that is designed for use in seawater. For example, installed in a pipeline on the seabed.

Swing check. Non-return valve that has a hinged disc as the closing element.

Swing check valve. A check valve with a single plate pivoted at the top and secured to the body of the valve. The flow of the process fluid pushes the plate open and in the event of flow reversal the plate swings to the closed position.

Tank valve. A valve arranged for fitting at the bottom of a tank or process vessel.

Tensile strength. The maximum amount of force that can be applied to a piece of material, before failure occurs. Also called the ultimate tensile strength, UTS.

Thermoplastic. A common term for plastic used for piping that loses strength as the temperature rises. Such plastic is used for utilities and fluids of a corrosive nature, usually operating at ambient temperatures.

Three-way valve. A diverter type valve that has three ports and allows the flow path of the process fluid to be switched, or two different flow paths to be combined.

Throttling. The regulation of the process fluid by positioning the closure element of the valve between open and closed to create the desire flow regime.

Through-conduit gate valve. A full-bore gate valve that has a very low-pressure drop and allows for the passage of pipeline pigs or scrapers for cleaning, de-watering, batching, etc.

Thrust. The force generated by any type of actuator to open, close, or position the closure element of a valve.

Top mounted handwheel. An accessory handwheel that is mounted on top of the actuator and used if there is a power failure.

Top works. Any number of parts that are located above the bonnet of the valve. They could be the yoke, the handwheel, the positioner, the actuator.

Torque. The rotational force applied to the shaft of a valve.

Toughness. A material's ability to remain undamaged when a force is applied. A tough material will deform first, before failure occurs.

Trim. The trim of the valve is the parts of the closure element that are exposed to the process flow, sometimes called the wetted parts.

Trunnion mounted ball valve. A robust ball valve, where the closure element ball is supported at the base by a shaft. This design is more common on larger valves and higher ratings, because of the weight of the ball.

Tubing. Small bore piping used to supply air or hydraulic fluid to an actuator.

Turbulence. A flow characteristic that is created when higher velocities and obstructions are experienced in a valve or a process system.

Ultrasonic testing. A testing method that requires the material to be bombarded with high frequencies to detect inclusions, pits, and cracks within the material. These reflected sound waves will find the depth at which the flaws occur.

Upstream. The process fluid before it reaches the valve.

Velocity. The speed at which the process fluid passes through the valve.

Vent. An opening in a piping system that can be exposed to the atmosphere and allows fluid to be released.

Viscosity. The resistance of a process fluid to flow. The "thickness" or "thinness." Highly viscous fluids (thicker) require more energy to move through a piping system.

Visual examination. Surface examination of a specimen that is carried out with the human eye without any supplementary test.

V-ring packing. A stem packing that is V shaped in cross section. Radial forces that are applied will force out the packing radially and create a tight seal against the wall of the packing box and the stem/shaft.

Wafer design. The construction of wafer design valves allows them to be "sandwiched" between flanged sections of pipeline. The benefit is lower bolting requirements. Typically used with certain butterfly and check valves.

Wall thickness. The thickness of the pressure-retaining shell of a valve. It must be designed to satisfy all the necessary tests that the valve will be subjected to during examination.

Waterhammer effect. The reaction when a valve is suddenly closed and a shock wave is transmitted through the piping system. This is generally caused by under sizing of the piping system. It is not only noise, but it can also cause mechanical damage to the piping system and associated equipment.

Weir. An obstruction in a diaphragm valve, against which the elastomer liner is compressed to prevent the flow of the process fluid.

Wellhead valve. Used to isolate the flow of oil or gas at the takeoff from an oil or gas well. The design is usually a plug or gate valve.

Yield strength. The force at which a material will begin to deform or stretch.

REFERENCES

[1] Lok, H. H. "Untersuchungen an Dichtungun Für Apparateflansche," Diss. Tech. High School, Delft, 1960.

[2] Weiner, R. S. "Basic Criteria and Definitions for Zero Fluid Leakage," Technical Report NO 32-926, *Jet Propulsion Laboratory Cal. Instn. of Techn.*, Pasadena, CA, December 1966.

[3] Wells, F. E. "A Survey of Leak Detection for Aerospace Hardware," paper presented at the National Fall Conference for the American Society of Nondestructive Testing, Detroit, October 1968.

[4] Selle, H. "Die Zündgefahren bei Verwendung verbrennlicher Schmier—und Dichtungsmittel für Sauerstoff-Hochdruck-armaturen," *J. Die Berufsgemeinschaft*, October 1951.

[5] Hill, R., E. H. Lee, and S. J. Tupper. "Method of Numerical Analysis of Plastic Flow in Plain Strain and Its Application to the Compression of a Ductile Material between Rough Plates," *J. Appl. Mech.*, June 1951, p. 49.

[6] F. H. Bielesch. SIGRI, GmbH, Meitingen, West Germany, Private Communications.

[7] "Spiral Wound Gaskets Design Criteria," Flexitallic Bulletin No. 171, Flexitallic Gasket Co., Inc., 1971.

[8] Krägeloh, E. "Die Wesentlichsten Prüfmethoden für It-Dichtungen," *J. Gummi und Asbest-Plastische Massen*, November 1955.

[9] *Compression Packings Handbook*, Fluid Sealing Association, 2017 Walnut Street, Philadelphia, PA.

[10] Morrison, J. B. "O-Rings and Interference Seals for Static Applications," *Machine Design*, February 7, 1957, pp. 91–94.

[11] Gillespie, L. H., D. O. Saxton and P. M. Chapmen. "New Design Data for FEP, TFE," *Machine Design*, January 21, 1960.

[12] Turnbull, D. E. "The Sealing Action of a Conventional Stuffing Box," *Brit. Hydr. Res. Assoc.*, RR 592, July 1958.

[13] Denny, D. F. and D. E. Turnbull. "Sealing Characteristic of Stuffing Box Seals for Rotating Shafts," *Proc. Instn. Mechn. Engrs.*, London, Vol. 174, No. 6, 1960.

[14] Rasmussen, L. M. "Corrosion by Valve Packing," *Corrosion*, Vol. 11, No. 4, 1955, pp. 25–40.

[15] Reynolds, H. J. Jr. "Mechanism of Corrosion or Stainless Steel Valve Stems by Packing—Methods of Prevention," Johns-Manville Prod. Corp. (Preprint No. 01-64 for issue by API Division Refining).

[16] Duffey, D. W. and E. A. Bake. "A Hermetically Sealed Valve for Nuclear Power Plant Service," V-Rep. 74-3, Edward Valves Inc.

[17] "Pressure Losses in Valves," *Engineering Sciences Data Item No. 69022*, Engineering Sciences Data Unit, London.

[18] ANSI/ISA 575.01, "Control Valve Sizing Equations."

[19] IEC Publication 543-1. "Industrial Control Valves, Part 1: General Considerations," International Electrotechnical Commission, Geneva, Switzerland.

[20] Hutchinson, J. W. "ISA Handbood of Control Valves," Instrument Society of America.

[21] Ball, J. W. "Cavitation Characteristics of Gate Valves and Globe Valves Used as Flow Regulators Under Heads up to about 125 Feet," *Trans. ASME*, Vol. 79, Paper No. 56-F-10, August 1957, pp. 1275–1283.

[22] Ball, J. W. "Sudden Enlargements in Pipelines," *Proc. ASCE Jour. Power Div.*, Vol. 88, No. P04, December 1962, pp. 15–27.

[23] Darvas, L. A. "Cavitation in Closed Conduit Flow Control Systems," *Civil Engg. Trans., Instn. Engrs. Aust.*, Vol. CE12, No. 2, October 1970, pp. 213–219.

[24] O'Brien, T. "Needle Valves with Abrupt Enlargements for the Control of a High Head Pipeline," *J. Instn. Engrs. Aust.*, Vol. 38, Nos. 10–11, October/November 1966, pp. 265–274.

[25] Streeter, V. L. and E. B. Wylie. "Fluid Transients." NY: McGraw-Hill Book Company.

[26] Kobori, T., S. Yokoyama and H. Miyashiro. "Propagation Velocity of Pressure Wave in Pipe Line," *Hitachi Hyoron*, Vol. 37, No. 10, October 1955.

[27] Randall, R. B., Brüel and Kjaer, Copenhagen, private communications.

[28] Ingard, U. "Attenuation and Regeneration of Sound in Ducts and Jet Diffusers," *J.Ac.Soc.Am.*, Vol. 31, No. 9, 1959, pp. 1202–1212.

[29] Allen, E. E. "Control Valve Noise," ISA Handbook of Control Valves, 1976.

[30] Boumann, H. D. "Universal Valve Noise Prediction Method," *ISA Handbook of Control Valves*, 1976.

[31] Arant, J. B. "Coping with Control Valve Noise," *ISA Handbook of Control Valves*, 1976.

[32] Scull, W. L. "Control Valve Noise Rating: Prediction Versus Reality," *ISA Handbook of Control Valves*, 1976.

[33] Sparks, C. R. and D. E. Lindgreen. "Design and Performance of High-Pressure Blowoff Silencers," *J. of Engg. for Industry*, May 1971.

[34] Bull, M. K. and D. C. Rennison. "Acoustic Radiation from Pipes with Internal Turbulent Gas Flows," proceedings from the Noise, Shock and Vibration Conference, Monash University, Melbourne, 1974, pp. 393–405.

[35] Bull, M. K. and M. P. Norton. "Effects of Internal Flow Disturbances on Acoustic Radiation from Pipes," proceedings from the Vibration and Noise Control Engineering Conference, Instn. Engrs., Sydney, 1976, pp. 61–65.

[36] Awtrey, P. H. "Pressure-Temperature Ratings of Steel Valves," *Heating/Piping/Air Conditioning*, May 1978, pp. 109–144.

[37] Champagne, R. P. "Study Sheds New Light on Whether Increased Packing Height Seals a Nuclear Valve Better," *Power*, May 1976.

[38] Cooper, Walter. "A Fresh Look at Spring Loaded Packing," *Chemical Engineering*, November 6, 1967, pp. 278–284.

[39] Häfele, C. H. Sempell Armaturen, Korschenbroich (Germany), private communications.

[40] Diederich, H. and V. Schwarz. *The Optimal Application of Butterfly Valves*, VAG-Armaturen GmbH, translation from *J. Schiff und Hafen*, No. 22, Seehafen-Verlag Erik Blumenfeld, Hamburg, August 1970, pp. 740–741.

[41] Holmberg, E. G. "Valve Design: Special for Corrosives," *Chem Engg.*, June 13, 1960.

[42] Holmberg, E. G. "Valves for Severe Corrosive Service," *Chem. Engg.*, June 27, 1960.

[43] Pool, E. B., A. J. Porwit, and J. L. Carlton. "Prediction of Surge Pressure from Check Valves for Nuclear Loops," ASME Publication 62-WA-219.

[44] Bernstein, M. D. and R. G. Friend. "ASME Code Safety Valve Rules—A Review and Discussion," *Trans. ASME*, Vol. 117, 1995, pp. 104–114.

[45] Papa, Donald M. "Back Pressure Considerations for Safety Relief Valves," Technical Seminar Paper, Anderson, Greenwood & Co.

[46] Huff, James E. "Intrinsic Back Pressure in Safety Valves," Paper presented at the 48th Midyear Refining Meeting of the American Petroleum Institute, Los Angeles, May 10, 1983.

[47] B. Föllmer (Bopp & Reuther). Reliable Operation of Type-Tested Safety Valves with the Influence of the Inlet Piping, 3R International, 31, No.7, July 1992.

[48] Teledyne Farris Safety and Relief Valves, Catalog No. FE-80-100, page 7.08.

[49] Walter W. Powell. "A Study of Resonant Phenonena in Pilot Operated Safety Relief Valves," Report No. 2-0175-51, Anderson, Greenwood & Co.

[50] Thompson, L. and O. E. Buxton Jr. "Maximum Isentropic Flow of Dry Saturated Steam through Pressure Relief Valves," *J. of Pressure Vessel Technology*, Vol. 101, May 1979, pp. 113–117, and ASME Publication PVP-33 "Safety Relief Valves," 1970, pp. 43–54.

INDEX